VOLUME ONE HUNDRED AND TWENTY FIVE

PROGRESS IN MOLECULAR BIOLOGY AND TRANSLATIONAL SCIENCE

Telomeres in Health and Disease

VOLUME ONE HUNDRED AND TWENTY FIVE

Progress in Molecular Biology and Translational Science

Telomeres in Health and Disease

Edited by

RODRIGO T. CALADO

Department of Internal Medicine, University of São Paulo at Ribeirão Preto School of Medicine, Ribeirão Preto, São Paulo, Brazil

ELSEVIER

AMSTERDAM • BOSTON • HEIDELBERG • LONDON
NEW YORK • OXFORD • PARIS • SAN DIEGO
SAN FRANCISCO • SINGAPORE • SYDNEY • TOKYO

Academic Press is an imprint of Elsevier

Academic Press is an imprint of Elsevier
225 Wyman Street, Waltham, MA 02451, USA
525 B Street, Suite 1800, San Diego, CA 92101-4495, USA
32 Jamestown Road, London NW1 7BY, UK
The Boulevard, Langford Lane, Kidlington, Oxford OX5 1GB, UK

First edition 2014

Notices

Knowledge and best practice in this field are constantly changing. As new research and experience broaden our understanding, changes in research methods, professional practices, or medical treatment may become necessary.

Practitioners and researchers must always rely on their own experience and knowledge in evaluating and using any information, methods, compounds, or experiments described herein. In using such information or methods they should be mindful of their own safety and the safety of others, including parties for whom they have a professional responsibility.

To the fullest extent of the law, neither the Publisher nor the authors, contributors, or editors, assume any liability for any injury and/or damage to persons or property as a matter of products liability, negligence or otherwise, or from any use or operation of any methods, products, instructions, or ideas contained in the material herein.

ISBN: 978-0-12-397898-1
ISSN: 1877-1173

For information on all Academic Press publications visit our website at store.elsevier.com

CONTENTS

CONTRIBUTORS

Geraldine Aubert
Terry Fox Laboratory, British Columbia Cancer Agency, Vancouver, British Columbia, Canada

Luis F.Z. Batista
Department of Medicine, and Department of Developmental Biology, Washington University School of Medicine, St. Louis, Missouri, USA

Clara Bodelon
Division of Cancer Epidemiology and Genetics, National Cancer Institute, National Institutes of Health, Bethesda, Maryland, USA

Rodrigo T. Calado
Department of Internal Medicine, University of São Paulo at Ribeirão Preto School of Medicine, Ribeirão Preto, São Paulo, Brazil

Maria Isabel Nogueira Cano
Depto. de Genética, Instituto de Biociências, Univ. Estadual Paulista Júlio de Mesquita Filho UNESP-Botucatu, São Paulo, Brazil

Marcelo Santos da Silva
Depto. de Genética, Instituto de Biociências, Univ. Estadual Paulista Júlio de Mesquita Filho UNESP-Botucatu, São Paulo, Brazil

Shahinaz M. Gadalla
Division of Cancer Epidemiology and Genetics, National Cancer Institute, National Institutes of Health, Bethesda, Maryland, USA

Miriam Aparecida Giardini
Depto. de Genética, Instituto de Biociências, Univ. Estadual Paulista Júlio de Mesquita Filho UNESP-Botucatu, São Paulo, Brazil

Daniel Hartmann
Department of General Surgery, Klinikum rechts der Isar, Technical University Munich, Munich, Germany

Norbert Hüser
Department of General Surgery, Klinikum rechts der Isar, Technical University Munich, Munich, Germany

Alexander Kleger
Department of Internal Medicine I, Ulm University Hospital, Ulm, Germany

Vinícius Santana Nunes
Depto. de Genética, Instituto de Biociências, Univ. Estadual Paulista Júlio de Mesquita Filho UNESP-Botucatu, São Paulo, Brazil

Raquel M.A. Paiva
Department of Internal Medicine, University of São Paulo at Ribeirão Preto School of Medicine, Ribeirão Preto, São Paulo, Brazil

Sharon A. Savage
Division of Cancer Epidemiology and Genetics, and National Institutes of Health, Bethesda, Maryland, USA

Marcela Segatto
Depto. de Genética, Instituto de Biociências, Univ. Estadual Paulista Júlio de Mesquita Filho UNESP-Botucatu, São Paulo, Brazil

Pavel Strnad
Department of Internal Medicine III and IZKF, University Hospital Aachen, Aachen, Germany

Yoshiaki Sunami
Department of General Surgery, Klinikum rechts der Isar, Technical University Munich, Munich, Germany

Guido von Figura
Department of Internal Medicine II, Klinikum rechts der Isar, Technical University Munich, Munich, Germany

PREFACE

The recognition of the role of telomeres in cell biology dates back to the 1930s, when two independent scientists recognized that the natural ends of linear chromosomes had specific properties. The contribution of telomere dysfunction to human disease, however, has much shorter history. Hermann Muller, while studying the genetic effects of radiation in *Drosophila*, observed that X-rays promoted chromosome breakage and fusion, but the natural ends of chromosomes (telomeres) were uniquely resistant to X-ray effects. He proposed that "the terminal gene must have a special function, that of sealing the end of the chromosome." He also hypothesized that "for some reason, a chromosome cannot persist indefinitely without having its ends thus sealed." He coined the term "telomere," from the Greek words *telo*, meaning "end," and *mere*, meaning "part." Barbara McClintock, working on a dicentric maize strain, observed that the ends of broken chromosomes were unstable and rapidly fused one another. However, if chromosome breaks happened in embryonic cells, the broken ends were able to "heal." We now know that this "healing" process is made possible by the addition of telomere sequence to the broken ends.

In 1971, the Russian theoretical biologist Alexey Olovnikov proposed that DNA polymerase was unable to fully replicate the ends of linear chromosomes, the "end-replication" problem or marginotomy. Given this enzymatic "deficiency," chromosomes are shorter after every cell division and Olovnikov hypothesized that telomeres functioned as chromosome "buffers," sparing important genetic material from erosion. Analyzing bacteriophages, James Watson came to similar conclusions in 1972. Elizabeth Blackburn and her colleagues Jack Szostak and Carol Greider solved the telomere problem in the 1980s. They identified the repetitive telomeric sequence and described a reverse transcriptase able to add nucleic acids to the 3′ ends of chromosomes, maintaining telomere lengths.

That telomere shortening causes disease was first observed in aplastic anemia, a bone marrow failure disorder in which the hematopoietic stem cell fails to produce adequate numbers of peripheral blood cells. Short telomeres in aplastic anemia impede the hematopoietic stem cell from proliferating and the marrow becomes aplastic. The first gene lesion in the telomerase complex was found in dyskeratosis congenita, an inherited type of aplastic anemia, in which the skin, lungs, and liver also are affected and patients have a proclivity for cancer development. Dyskeratosis congenita became central in telomere biology and dysfunction, as novel components of the telomerase complex and telomere biology genes were discovered in affected pedigrees. Also important, dyskeratosis congenita and telomere dysfunction molecularly linked illnesses that were otherwise unrelated. Aplastic anemia, idiopathic pulmonary fibrosis, liver cirrhosis, and acute myeloid leukemia are diseases with specific clinical presentations, treatments, and prognosis, but in some cases they share the same basic etiology: excessive telomere erosion. The identification of this mechanism makes possible a new disease classification, based not on clinical phenotype or affected organs but rather on a common molecular event. This novel way of interpreting diseases may be mainly beneficial for the discovery of novel therapies.

The goal of this book is to discuss the importance of telomeres in human physiology and disease, putting together illnesses that until now were unrelated and far apart in internal medicine textbooks.

I would like to truly thank the contributing authors of this book for their time, expertise, and for making this volume novel, educational, and informative. Finally, I would like to express my gratitude to Dr. Michael Conn, Chief Editor of the *Progress in Molecular Biology and Translational Science* series, and Mary Ann Zimmerman and Helene Kabes from Elsevier for their continuous help.

Rodrigo T. Calado

CHAPTER ONE

Telomere and Telomerase Biology

Miriam Aparecida Giardini, Marcela Segatto, Marcelo Santos da Silva, Vinícius Santana Nunes, Maria Isabel Nogueira Cano
Depto. de Genética, Instituto de Biociências, Univ. Estadual Paulista Júlio de Mesquita Filho UNESP-Botucatu, São Paulo, Brazil

Contents

Abstract

Telomeres are the physical ends of eukaryotic linear chromosomes. Telomeres form special structures that cap chromosome ends to prevent degradation by nucleolytic attack and to distinguish chromosome termini from DNA double-strand breaks. With few exceptions, telomeres are composed primarily of repetitive DNA associated with proteins that interact specifically with double- or single-stranded telomeric DNA or with each other, forming highly ordered and dynamic complexes involved in telomere maintenance and length regulation. In proliferative cells and unicellular organisms, telomeric DNA is replicated by the actions of telomerase, a specialized reverse transcriptase. In the absence of telomerase, some cells employ a recombination-based DNA replication pathway known as alternative lengthening of telomeres. However, mammalian somatic cells that naturally lack telomerase activity show telomere shortening with increasing age leading to cell cycle arrest and senescence. In another way, mutations or deletions

Progress in Molecular Biology and Translational Science, Volume 125
ISSN 1877-1173
http://dx.doi.org/10.1016/B978-0-12-397898-1.00001-3

of telomerase components can lead to inherited genetic disorders, and the depletion of telomeric proteins can elicit the action of distinct kinases-dependent DNA damage response, culminating in chromosomal abnormalities that are incompatible with life. In addition to the intricate network formed by the interrelationships among telomeric proteins, long noncoding RNAs that arise from subtelomeric regions, named telomeric repeat-containing RNA, are also implicated in telomerase regulation and telomere maintenance. The goal for the next years is to increase our knowledge about the mechanisms that regulate telomere homeostasis and the means by which their absence or defect can elicit telomere dysfunction, which generally results in gross genomic instability and genetic diseases.

1. INTRODUCTION

Telomeres are stretches of tandemly repeated and, in most cases, conserved DNA sequences that form the physical ends of eukaryotic chromosomes.[1–4] Since the discovery of telomerase in the mid-1980s, telomere biology has attained great interest from the scientific community due to a consensus concerning the importance of these structures for genome stability and cell proliferation.[5–11] Briefly, telomeres cap chromosome termini to distinguish these natural ends from double-stranded DNA (dsDNA) breaks and to avoid fusions of telomeric sequences.[12,13] How do they do that? Telomeres serve as substrates for telomerase, the enzyme responsible for adding DNA to the ends of chromosomes, thus maintaining chromosome length.[14,15] Telomerase is a specialized reverse transcriptase ribonucleoprotein (RNP) composed of two main components, a telomerase reverse transcriptase (TERT) protein and a noncoding RNA component (TER, telomerase RNA), which is an integral and essential part of the enzyme. TER contains a short template sequence that is copied by telomerase during telomere replication.[16,17] A minimal complex formed by TERT and TER is sufficient for *in vitro* enzyme activity. However, *in vivo*, enzyme biogenesis, enzyme activity, and nucleotide addition processivity are also dependent on other accessory proteins, indicating that a relatively complex maturation pathway is likely required to generate the active RNP, which must subsequently find its substrate.[16,17] Telomerase works to circumvent the loss of terminal DNA, which is caused, in part, by the inability of DNA polymerases to completely replicate the 5′ ends of linear DNA molecules and by the actions of exonucleases. Both processes generate transient 3′ G-overhangs on the opposite ends of both the leading and lagging strands. These overhangs are recognized by end-binding proteins, which bind to the overhangs and subsequently recruit telomerase to elongate the G-strand termini. Once the telomeres are replicated by telomerase, the C-strand is synthesized by the conventional DNA replication machinery.[5,18–22]

Depending on the organism, the expression of some of the telomerase holoenzyme components is strictly regulated.[23] For example, TERT and TER are constitutively expressed in unicellular eukaryotes, whereas in mammals, TERT is expressed only in highly proliferative cells and tumor cells. Enzyme activity is completely absent in somatic cells and in cells with low proliferative capacity, which explains why telomerase is being extensively explored as a potential target for antitumor therapy.[15,24] The extra-telomeric functions of telomerase also indicate that this enzyme can be considered an important factor against replicative senescence.[25–27] However, telomerase is not the only mechanism used to maintain chromosome length; in some organisms, such as the fruit fly *Drosophila melanogaster*, retrotransposon-like elements are alternatively used to replenish the DNA at the ends of chromosomes.[28] In addition, under certain circumstances, yeast and human cells that lack telomerase activity, as well as some telomerase-negative tumor lineages, are able to maintain telomeres using a recombination-based DNA replication mechanism known as alternative lengthening of telomeres (ALT).[29–31]

However, telomeres are not only formed by repetitive DNA sequences; they are nucleoprotein structures formed by interactions between proteins that, most of the time, specifically recognize double-stranded and G-rich single-stranded DNA. The actions of these proteins at telomeres are dynamically orchestrated by other protein factors that are not necessarily exclusive to telomeres. These proteins usually interact with telomeric proteins or chemically modify them, helping to cap the ends and to control telomerase activity, consequently regulating telomere length.[32,33] In addition, it was recently shown that telomeres can be controlled by long, noncoding telomeric RNA transcripts known as telomeric repeat-containing RNAs (TERRAs), which originate from the subtelomeres near the end of the C-rich strand. These transcripts contain, near their 3′ end, telomeric repeats in the form of RNA.[34] It is generally accepted that TERRA is transcribed from short telomeres and that its transcription *in vivo* likely recruits telomerase molecules to elongate short telomeres.[35]

In contrast, telomere deprotection can be disastrous for cell survival and life span because it usually results in DNA damage response, which can lead to senescence or apoptosis and genome instability.[36] Telomeres can become deprotected as a result of shortening, which is caused mainly by a lack of telomerase activity or a deficiency in telomerase or a telomeric chromatin component, affecting development and disease in mammals.[37]

This chapter summarizes the main, sophisticated regulatory circuitries responsible for maintaining functional telomeres, with a focus on the principal mechanisms responsible for telomere replication and length regulation.

2. THE GENERAL STRUCTURE OF TELOMERIC DNA

Most eukaryotic species, ranging from primitive to higher species, contain tracts of repetitive DNA at the ends of linear chromosomes. These tracts are mainly composed of subtelomeric and telomeric sequences with both conserved and divergent features. Telomeric DNA is characterized by G-rich repetitive sequences, such as the vertebrate hexanucleotide repeat TTAGGG.[2,15,38] The strand that forms the 3′ end of the chromosome and protrudes toward the chromosome terminus is usually G-rich and longer than the opposite C-rich strand. The G-rich strand forms a 3′ single-stranded overhang, which is found at most eukaryotic chromosome termini[4,39,40] (Fig. 1.1A). This overhang can be formed by three processes: the degradation of the RNA primer from the 5′ end of the newly synthesized strand

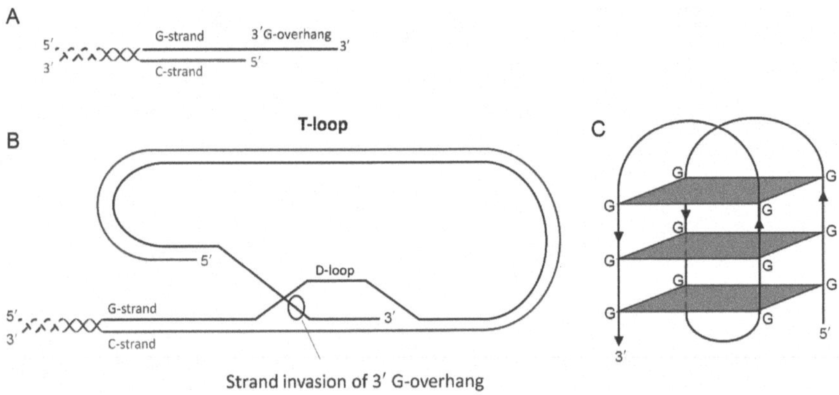

Figure 1.1 Different telomere configurations. (A) Telomeric DNA is formed of double-stranded and single-stranded DNA. One of the strands is C-rich, and the opposite G-strand protrudes toward the end of the chromosome to form a 3′ G-overhang. (B) A telomere forming a T-loop, a lariat-like configuration that arises by strand invasion of the telomeric 3′ G-overhang into the upstream telomeric double-stranded DNA, forming an internal D-loop. (C) G-quadruplex DNA is another specialized structure formed at the ends of chromosomes. This configuration is formed from G-quartets, which are square, planar arrays of four guanines (Gs) that are hydrogen-bonded by Hoogsteen base pairing. *Adapted from Nandakumar and Cech.*[41]

during semiconservative DNA replication; the 3′ end elongation mediated by telomerase, followed by C-strand fill-in synthesis; and the end resection of both chromosome ends by nucleolytic processing.[21,40] Toward the chromosome center and directly adjacent to telomeres lie the subtelomeric sequences, which represent highly polymorphic chromosome regions composed of a combination of various conserved and nonconserved repeats in virtually all studied eukaryotes. In some pathogens, for example, subtelomeres are highly recombinogenic sites that serve as the location of genes encoding the surface antigens used in host immune evasion.[3,42]

In general, telomeres show high sequence variation,[43] although most telomeric sequences are clearly related to each other (e.g., mammals and trypanosomes share the same TTAGGG telomeric repeats) and consist of telomeric DNA formed by simple, tandemly repeated sequences. The exceptions are telomeres in many yeast species, which exhibit short, complex sequences, and telomeres in some flies and plants species, which can be organized as retrotransposable elements.[44,45] The generation of regular or irregular telomeric sequences appears to be directly dependent on the action of telomerase. For example, although common telomerases faithfully copy one telomeric repeat of the TER template sequence, error-prone telomerases, such as the telomerases of some ciliate protozoa, alter their TER template usage, that is, they misincorporate nucleotides during synthesis.[46,47] In addition, some yeast species exhibit low-fidelity telomerase enzymes, which copy different parts of the TER template, resulting in telomeres containing a mix of variant repeats.[48,49] Moreover, the absence of telomerase activity in *Drosophila* spp., for example, is compensated by the action of three retrotransposable elements organized as long tandem head-to-tail arrays at the end of the chromosome.[28]

A common feature shared by almost all eukaryotic telomeric DNA is heterogeneity in the length of telomeric repeats at individual chromosome ends; in mammals, this is predominantly observed in somatic, but not germline, cells.[43] The length of the telomeric repeat tracts can also vary greatly between species and within the same species. For example, mouse telomeres (~100 kb) are much longer than human telomeres (ranging from ~10 to 15 kb).[2,50]

All these observations suggest that telomeres have a common evolutionary origin.[38] Several controversies exist regarding this issue. Some authors argue that telomeres originally descended from an ancient replication-competent retrovirus,[51] whereas others emphasize that the primordial terminal structures of eukaryotic linear chromosomes are derived from selfish element(s) (e.g., retrotransposable elements), which caused the linearization of ancestral circular genomes.[52] Nevertheless, regardless of their origin, the acquisition of telomere repeats was a key event in the evolution of the eukaryotic nucleus, where telomeres play crucial roles in chromosome organization, genome stability, and end protection.

2.1. Telomere loops and G-quartets: Specialized structures at the ends of eukaryotic chromosomes

Telomeres from distinct organisms can form higher-order structures maintained by telomeric proteins.[53,54] Telomere loops (T-loops) are an example of these structures. T-loops exhibit a lariat-like configuration that arises by the strand invasion of a 3′ single-stranded DNA overhang (approximately 100–200 nucleotides of TTAGGG repeats) into upstream telomeric dsDNA, forming a displacement (D) loop of TTAGGG repeats called a D-loop (Fig. 1.1B).[54] The existence of T-loops has primarily been demonstrated in humans using electron microscopy to analyze photo-cross-linked telomeric chromatin.[53] Later, T-loops were also described in trypanosomes and fission yeast telomeres.[53–55] Electron microscopic studies of uncross-linked telomeric chromatin from chicken erythrocytes and mouse lymphocytes have revealed similar structures.[56] T-loops can be very large (hundreds of kilobases). In some cases, T-loops encompass the entire telomere, although the exact site of the 3′ single-stranded DNA invasion has not been established.[41,54] There is a close correlation between the length of the telomeric repeat array and the size of the T-loops. However, it remains unclear why T-loops always extend to the "beginning" of the telomere (the junction between the telomeric and subtelomeric DNAs) in primary cells and how their lengths vary according to the age of the cell.[56] Because T-loops appear to be a highly protected and inert structure, more research is necessary to understand the mechanism underlying how they are opened to allow telomerase access to the end of the chromosome. It is worth noting that T-loop formation is thermodynamically unfavorable; therefore, it is not surprising that protein factors other than the conserved telomeric components have been implicated in T-loop assembly.[41] It is also worth mentioning that under certain circumstances, T-loops may be converted to or may generate (e.g., by recombination) extrachromosomal T-circles, which are sometimes associated with the generation of telomeres in the absence of telomerase, a phenomenon known as ALT (discussed in a later section of this chapter).[30,57]

Another specialized structure formed at the ends of telomeres is G-quadruplex DNA, also known as G-tetrads or G_4-DNA. This structure is composed of G-quartets (also known as guanine tetrads), which are square planar arrays of four guanines (Gs) that are hydrogen-bonded by Hoogsteen base pairing.[41] Two or more G-quartets can stack on top of each other to form a G-quadruplex (Fig. 1.1C). Intramolecular G-quadruplexes form spontaneously *in vitro* because telomeric DNA sequences contain at least four blocks of Gs, which form the G-quadruplex structure only when the appropriate ionic conditions are present. G-quadruplex formation requires a thermodynamic prerequisite; thus, some telomeric proteins help to catalyze G-quadruplex folding and unfolding.[58,59] All of the known telomeric-capping proteins bind to unfolded telomeric DNA and ignore the G-quadruplex, which provides a strong argument that most telomeric DNAs in cells are not folded into quadruplexes.[41,60] Nevertheless, these structures can also occur *in vivo*, and they possibly provide an alternative means of telomere protection when telomeric proteins have been displaced.[53]

The greatest challenge to an improved understanding of the functions of G-quadruplexes and T-loop DNA is the difficulty in effectively inactivating these structures to result in a specific cell phenotype. This ability could certainly be useful, for instance, to limit telomerase extension in cancer cells by G-quadruplex-stabilizing agents.[61,62]

3. REPLICATION OF TELOMERES

The first descriptions of the importance of telomeres in protecting chromosome ends date to the late 1930s, when Muller and McClintock[63,64] studied fly and corn chromosomes, respectively, exposed to high doses of X-rays. Both researchers noted that ionizing irradiation induced DNA breaks along the chromosomes. However, at the ends, the chromosomes did not exhibit breaks or deletions; instead, they fused to each other or to broken chromosomes. Muller[64] introduced the term "telomere" (derived from two Greek words: telos (terminus) and meros (part)) for these special structures, which, according him, exhibited a special heterochromatic morphology and were able to "cap" the ends of chromosomes.

Ten years later, the discovery of DNA polymerases and their ability to synthesize DNA in only one direction (5′–3′), which prevents the finalization of DNA replication at the chromosome ends, led Olovnikov and Watson to propose independent theories referring to an "end-replication problem".[18,20] Olovnikov went further with his theory, also known as marginotomy, suggesting that if these ends were not replenished (e.g., in somatic cells, which have a limited life span), a gradual shortening of chromosome ends would occur. After many rounds of cell division, this shortening would lead to the loss of essential genes and, consequently, cellular senescence. Moreover, Olovnikov brilliantly proposed that the length of the terminal sequences could determine the possible number of DNA replication rounds.

The sequencing of the native telomeres of a ciliate protozoon by Blackburn in 1978[1] showed that these structures were mainly composed of short, tandemly repeated DNA sequences. Later, with more knowledge about telomeres from lower to higher eukaryotes, a consensus grew about their structural conservation: most telomeres, with few exceptions, exhibit and share some common features.[3,65,66]

However, until the beginning of the 1980s, the intriguing question of how telomeres are replicated remained unanswered. Hints about the existence of a specialized enzyme able to replenish chromosome ends came with the results of experiments showing that native linear DNAs from ciliate protozoa introduced into yeast were maintained for generations by the addition of yeast 5′-TG_{1-3}-3′ telomeric DNA, instead of the ciliate T2G4 or T4G4 telomeric repeat.[39,67] Almost simultaneously, Bernards *et al.*[68] reported that telomeres of the pathogenic protozoa *Trypanosoma brucei* grew gradually during mammalian infection. Altogether, these results indicated the existence of a mechanism able to maintain chromosome ends that was conserved among different eukaryotes.

In the mid-1980s, Carol Greider (in Liz Blackburn's laboratory) analyzed protein extracts from the ciliate *Tetrahymena thermophila* after mating and discovered an enzyme activity that elongated telomeres, which they named telomerase.[5,69] This enzyme not only behaved similarly to a reverse transcriptase but also contained conserved motifs for the reverse transcription of a template sequence present within an RNA component subunit (the TER component), which is used for the addition of G-rich telomeric repeats to the 3′ single-stranded ends of chromosomes.[19,70,71]

Actually, many lines of evidence show that telomere replication is a multistep process that requires dynamic interactions among multiple factors comprising components of the telomerase holoenzyme and different proteins that form telomeric chromatin.

3.1. The telomerase RNP complex

Telomerase is a high-molecular weight RNP complex that consists of two major components: TER and a TERT (Fig. 1.2A). TER is a noncoding RNA that is essential for telomere synthesis; it serves as a template to

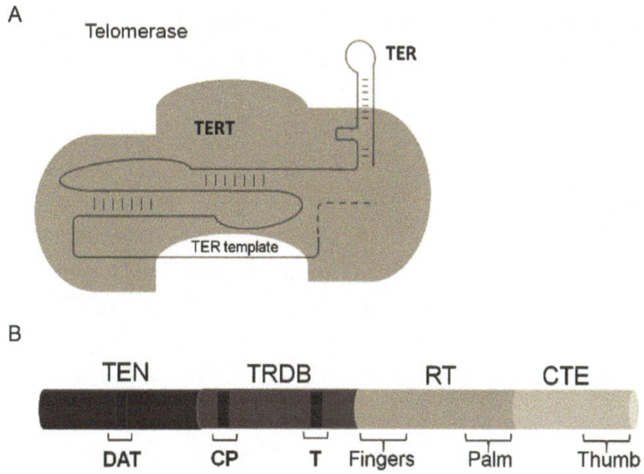

Figure 1.2 *Telomerase holoenzyme.* (A) The telomerase minimal complex formed by the reverse transcriptase component (TERT) and the RNA component (TER). The position of the TER template domain is signaled. (B) Schematic representation of the TERT primary structure. The most important TERT domains are indicated. TEN, telomerase N-terminal domain; TRBD, telomerase RNA-binding domain; RT, reverse transcriptase domain; CTE, C-terminal extension. The position of the structural fingers, palm, and thumb subdomains is highlighted. *Panel (B): Adapted from Nandakumar and Cech.*[41]

elongate the 3′ overhang of the telomeric G-rich strand.[5,19] TERT is a protein component that acts as a specialized reverse transcriptase and contains conserved catalytic domains. In humans, only the TERT and TER components are necessary to ensure telomerase activity *in vitro*, although some regulatory proteins associated with the holoenzyme complex are also essential for the catalytic function of telomerase *in vivo*.[72] It is estimated that at least 32 different proteins associate with human telomerase *in vivo* to maintain its functionality[73]; only some of these proteins are phylogenetically conserved.

In general, telomerase RNP complexes exhibit conserved compositions and structures, even in evolutionarily distant organisms. Their compositions are similar from yeasts to mammals, including humans.[74–76] Some of the most important components of this holoenzyme are described below.

3.1.1 The reverse transcriptase component

In addition to the RNA subunit (TER), the telomerase core enzyme contains the TERT reverse transcriptase subunit,[77,78] which is now known to have several functions.

The TERT component contains several telomerase-specific motifs and species-specific linkers, as well as all the reverse transcriptase motifs commonly found in non-LTR (long terminal repeat) reverse transcriptases. TERT components vary in size substantially, but most of the domains and their distribution are conserved among even phylogenetically distinct organisms.[79] The TERT molecular weight range is approximately 1000–1500 amino acids.[71,77,80] On the basis of its structure and function, the TERT polypeptide can be subdivided into three major domains: the telomerase essential N-terminal (TEN) domain, the TERT RNA-binding domain (TRBD), and the reverse transcriptase domain, which contains the active site for reverse transcription. The reverse transcriptase domain comprises motifs 1, 2, and A–E, which represent the fingers and palm of a hand, and a C-terminal extension, which represents the thumb subdomain (also known as the less-conserved protein region)[71,77] (Fig. 1.2B).

The TEN domain of TERT has been implicated in providing an "anchor site" that binds to telomeric DNA upstream of the primer–template interaction.[81,82] This domain is highly conserved among vertebrate proteins, which is an exception to the generally low level of sequence conservation beyond the active site. The crystal structure of the isolated TEN domain of the *T. thermophila* TERT revealed a novel protein fold with a Trp187 within the groove, which suggests that the surface encompassing this residue is involved in DNA binding.[83] A similar anchor site resides within the TEN domain of the human TERT.[84,85] In addition, the human TEN domain contains a DAT motif,[86] which has been implicated in telomerase recruitment, whereas the yeast TEN domain contributes to the binding of the Est3 accessory subunit.[87]

The crystal structures both of a putative TERT from the beetle *Tribolium castaneum* and of its catalytic subunit cocrystallized with an RNA–DNA hairpin designed to resemble the putative RNA template region and telomeric DNA[88,89] revealed close contacts between the TRBD and the thumb subdomain of the reverse transcriptase. This interaction resulted in a closed ring-like tertiary structure with a large cavity at its center, which appeared sufficiently large to bind the primer–template duplex.[88] The *T. castaneum* crystallographic structure showed that the fingers and palm of TERT interact with the backbone of the RNA to place the template in the active site, whereas the T-pocket (containing the T-motif residues) and the ciliated protozoan pocket (containing the CP motif) bind to a region upstream of the template that presumably mimics the template boundary element of TER. Although the fingers and palm hold the RNA template, the thumb of the reverse transcriptase binds and secures the DNA primer.[89]

In addition to providing the active site for catalysis, the reverse transcriptase motifs secure the RNA component to the protein, ensuring the maintenance of a stable RNP while allowing the template to move through the active site.[90] Movement is essential; a single active site must accommodate the addition of multiple nucleotides, after which the translocation of the template relative to the DNA product is necessary for multiple rounds of nucleotide addition[91] (Fig. 1.3).

The biogenesis of the telomerase RNP proceeds through sequential steps involving protein–RNA interactions. Budding yeast, vertebrates, and ciliates use different TER-binding proteins to produce a biologically stable telomerase RNP. The assembly of RNP with TERT then leads to RNP activation, which appears to be a dynamic process. Additional proteins give the holoenzyme the ability to function on a chromosome substrate. In addition to TERT, the TER component and the single-stranded DNA form a network of protein–nucleic acid interactions, which orchestrate the proper positioning of the template and primer in the active telomerase site.[92]

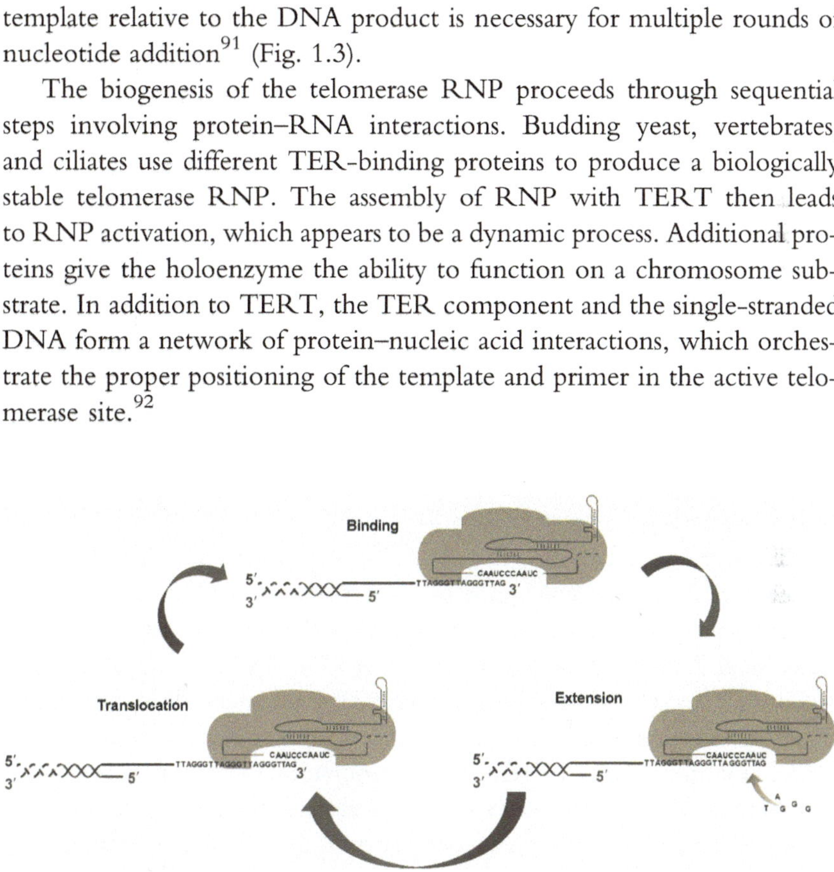

Figure 1.3 *Telomere elongation by telomerase.* The telomeric 3′ G-overhang binds to the TER template, while nucleotides upstream of the telomeric DNA interact with the TERT anchor site (*Binding*). Next, TERT reverse-transcribes one copy of the telomeric repeat from the TER template sequence by adding nucleotides onto the 3′ end of the telomeric DNA until the 5′ end of the template is encountered (*Extension*). This process is known as nucleotide addition processivity (NAP). Then, the TERT active site translocates to reposition itself and the 3′ end of the template at the newly formed telomeric repeat (*Translocation*). Another round of nucleotide addition is then initiated. *Adapted from Autexier and Lue.*[23]

3.1.2 Telomerase accessory factors

In addition to the catalytic core of telomerase, which is composed of TERT and TER, several telomerase accessory factors assist in mammalian and yeast telomerase assembly, maturation, recruitment, and activation.[17,92] Here, we review the protein classes and complexes that participate in these processes from yeast and mammalian telomeres.

Est1p (ever short telomeres 1) was first identified in yeast mutants carrying very short telomeres and was later found to be associated with yeast telomerase.[93] Est1p is essential for enzyme activity *in vivo*, but it is dispensable in *in vitro* assays. Est1p interacts directly with Cdc13, the major yeast telomere end-binding protein (TEBP; see section 4). This interaction is important for recruiting telomerase to the chromosome ends[94] and for somehow activating telomerase already associated with telomeres.[95] A human Est1 ortholog, EST1A, is associated with most or all of the active telomerase in human protein extracts and is involved in chromosome end capping and telomere elongation.[96,97] Another yeast subunit, Est3p, is also important for enzyme activity *in vivo* but not *in vitro*, although its specific function remains unknown.[91] The AAA+ATPases pontin and reptin, which are involved in many other cellular functions, play direct roles in mammalian telomerase RNP biogenesis and enzyme activity *in vivo*.[98] Other proteins that are involved in telomerase biogenesis are directly associated with the TER component (see next section for details).

Transient associations of TER and TERT with chaperone activities have also been reported and apparently occur during the assembly, disassembly, and degradation of telomerase complexes. Based on well-established precedents for the assembly of Sm proteins and H/ACA-motif RNA-binding proteins,[99,100] the pathways for telomerase RNP biogenesis in budding yeast and vertebrate cells must involve the participation of RNP assembly chaperones. The heat-shock protein 90 (HSP90) and p23 chaperones associate with human TERT, and this association is blocked by the HSP90 inhibitor geldanamycin, which reduces the activity of both recombinant and endogenous telomerases *in vitro*.[101,102]

3.1.3 The telomerase RNA component

The TER component provides the template sequence for reverse transcription and helps to assemble the RNP complex during its maturation process.[103,104] The interaction between TER and the protein component TERT determines the catalytic activity, processivity, and telomere-binding ability of telomerase.[105–107]

The mechanism underlying telomerase function is largely conserved among different organisms; however, both the protein and RNA components vary greatly in length and sequence conservation. Compared with TERT, the TER component shows greater variability in size and primary sequence, with sizes ranging from 147–209 nt in ciliated protozoa to approximately 450 nt in mammals and 2.2 kb in *Plasmodium falciparum* and *T. brucei*.[108–112] It has been suggested that this TER variability may be due to a freedom to drift or to the intrinsic ability of RNA to change or adapt to the environment to improve its function and the enzymatic performance of RNP.[110]

The yeast and mammalian TER components share similarities: they both have a 5′-trimethyl cap structure. In contrast, whereas the yeast TER is more similar to small nuclear RNAs because it is associated with the SM7 complex at its nonpolyadenylated 3′ end, the mammalian TER is considered a small nucleolar RNA, and its 3′end is associated with dyskerin and the TCAB1 (telomerase Cajal body protein 1) protein. In all cases, these proteins are directly involved in the biogenesis and nuclear retention of telomerase.[17]

Despite exhibiting different sizes and no or very little similarity in nucleotide sequence, even among very close organisms, some elements of the TER secondary structure play the same functional roles in ciliates, yeasts, and vertebrates, suggesting that the basic biochemical activities of TER are conserved in these organisms.[108,113] These common motifs are shared by all known TERs and are involved in RNA stabilization, accumulation, subcellular localization, and assembly with other components.[114] The template/pseudoknot domain[108,115] is present in all TERs, while a second domain can be a stem–loop IV in ciliate TERs, a bulged three-way stem junction in budding yeast, or a combination of both, as in the vertebrate CR4/5 motif.[110,116–118] Indeed, it has been demonstrated that when isolated from TER and combined *in trans* with TERT, the template/pseudoknot and CR4/5 domains are sufficient to restore human telomerase activity *in vitro*.[119,120] In vertebrates, there is an additional H/ACA-type domain at the 3′ end that allows for the binding of four accessory proteins—dyskerin, NOP10, NHP2, and GAR1—that are necessary for telomerase biogenesis and localization *in vivo*.[114,121] In agreement with this finding, a large variety of proteins are found tightly associated with TER and require the assembly and maturation of the telomerase RNP. These associations are much more evolutionarily divergent than the association of TERT and its interacting factors. These proteins are mandatory subunits of an endogenously assembled telomerase RNP, and they constitute the structural core of all telomerase holoenzymes. The two subunits of the KU heterodimer comprise one class of these telomerase proteins in yeast. KU is responsible for the nonhomologous end joining (NHEJ) of broken chromosomes, and its presence in the telomerase holoenzyme was initially a surprise. It was further demonstrated that KU binds directly to TER (in yeast, TER is also called TLC1) and promotes the *de novo* addition of telomeres to chromosome ends, thereby helping to promote telomere healing.[122] It is now apparent that the primary role of KU in facilitating telomerase recruitment is at the level of TER nuclear retention.[123] The dissociation of these proteins from TER is expected to induce RNP turnover.[124]

In addition, mutations in the TER primary sequence can lead to a different secondary structure, which affects telomerase function by disrupting interactions with TERT or accessory proteins, changing template positioning, and thus decreasing telomerase activity in cells. Deficiencies in TER structure have been linked to several diseases in humans.[125–128]

The importance of the TER component for telomerase function *in vivo* has been extensively studied in mice, in which TER gene disruption abolishes telomerase activity and shortens telomeres,[129] altering the long-term viability of highly proliferative tissues and compromising genomic integrity.[130] In addition, short telomeres have been shown to suppress tumor progression, reducing oncogenesis in subsequent generations.[131,132] However, when telomerase was reintroduced in these mice, it recognized and elongated the short telomeres, preventing end-to-end fusions and senescence phenotypes and restoring the oncogenic potential.[131,133,134] These data suggest the possible use of telomerase as gene therapy for human diseases that cause premature aging or other diseases related to the normal aging process and as a potential target for the development of new drugs that block tumor growth.[132]

3.1.4 How telomeres are replicated by telomerase

Most telomeric DNA is replicated by semiconservative DNA replication. Therefore, the two telomeres generated by the replication of the end of each chromosome arm are synthesized differently. One telomere is produced through leading-strand synthesis, whereas the other is the product of lagging-strand synthesis. Whereas the leading-strand telomere is expected to be blunt ended, the removal of the final RNA primer from the telomeric DNA strands replicated by lagging-strand synthesis can result in the loss of a small amount of telomeric DNA at each round of DNA replication, in addition to the resection of both telomere ends by exonucleases.[21] To counteract this continuous loss of DNA, nearly all eukaryotes, with the notable exceptions of some flies and plants, use the telomerase enzyme to lengthen telomeric DNA.[135]

The major activity of telomerase is to ensure RNA-dependent telomere elongation.[23] The telomerase catalytic cycle consists of several sequential stages. One telomeric repeat is added after substrate binding. The resulting product can either dissociate from the enzyme's active site or undergo translocation, followed by elongation (Fig. 1.3). The ability of telomerase to move the synthesized DNA to the template's start site indicates that two processivity types are involved in its function. Nucleotide addition (type I processivity) is intrinsic to all polymerases because repeat addition (type II processivity) is unique to telomerase and determines the ability of an enzyme to repeatedly copy an RNA template region via the elongation of a single substrate molecule.[23,136,137]

Primer binding at the first stage of the telomerase reaction cycle is required by its complementary action with the TER template region. When primers with different sequences are used, the efficiency of the formation of the complex with an enzyme does not correlate with the length of the resulting DNA–RNA duplex[138] because telomerase binds to the substrate immediately after the participation of the template region of TER. Structural elements of the TERT active site regulate the efficiency of duplex formation and the translocation of the freshly synthesized product during the processive synthesis of telomeric repeats. The anchor regions of TERT and TER also participate in the primary binding of the primer. Nucleotides bind to the primer during the second stage of the telomerase reaction cycle.[23,139] The TER template region constitutes approximately 1.5 telomeric repeats, which are first positioned on and annealed to the TERT TRBD domain to enable the enzyme to produce perfectly homogeneous tandem repeats by copying one telomeric repeat of the template region to the chromosome 3′ overhangs.[23]

The major feature of telomerase is its ability to processively add repeats.[137] The mechanism of telomerase translocation after a repeat is synthesized remains unknown. Whether enzyme processivity of this type is required for efficient telomere elongation remains an open question. It was previously ascertained that critically short telomeres elongate processively.[140] A set of products with different numbers of telomeric repeats are formed during telomerase operation. After a single telomeric repeat is added, either the reaction is terminated or the rate of reaction decreases; in other words, template translocation and annealing represent the rate-limiting stages. It has been demonstrated that shelterin components, the POT1 and TPP1 proteins, form a complex that efficiently stimulates telomerase activity and processivity.[141] Telomerase activity is inhibited when the complex is bound to the 3′ terminus of the primer, but it functions processively when it is bound to the 5′ terminus. The TIN2 protein, another component of the telomere-associated shelterin complex (see below), also plays a role in promoting telomerase action.[41] Very recently, it was shown that the mammalian CST end-binding telomeric complex can be considered a terminator of telomerase action and, thus, of telomere replication. It appears that the elongation of the telomeric 3′ G-overhang by telomerase increases the binding of CST to the telomere's 3′ end, leading to telomerase inhibition and the recruitment of DNA polymerase-α to perform C-strand fill-in synthesis[142] (see below for details about the CST complex activity).

Moreover, it was shown that telomerase does not act on each telomere during each cell cycle; instead, it has a preference for short telomeres. This finding demonstrates that short telomeres are elongated more frequently than long telomeres, and they can alternate between extendable and non-extendable states.[143]

3.1.5 Regulation of telomerase activity in mammals

Due to the important role of telomerase activity in the maintenance of genome integrity, telomerase is extensively regulated. Multicellular organisms exhibit developmental, tissue-specific, and stress-responsive strategies for telomerase repression.[144,145] The inactivation of telomerase and the maintenance of telomere length in human somatic cells have been proposed to function as a tumor suppressor mechanism.[146,147] The inactivation of telomerase may also be required for quiescence, differentiation, and the death of some cell types.[148] However, cumulative telomere erosion limits the renewal capacity of highly proliferative human cell lineages in the skin and blood.[92]

Whereas TER expression is ubiquitous, TERT expression appears to be highly regulated in some organisms, principally in mammals. Several strategies have been proposed to control telomerase activity, because this multisubunit enzyme can be regulated at various levels, including expression control. For example, the epigenetic modification of histones can modulate chromatin structure and the accessibility of the transcriptional machinery to regulatory regions of target genes. In this context, numerous transcription factors, such as c-MYC, SP1, MAD1, and HIF-2a, have been shown to recruit either histone acetyltransferases or histone deacetylases to the TERT promoter to control TERT expression.[149–151] However, the transcription expression is not always correlated with the enzyme activity, which might result in transcription modulation failure.[152] Accordingly, telomerase is expressed in stem cell compartments and in embryonic stem cells, but TERT expression and telomerase activity are often very low or undetectable in somatic cells.[153] In contrast, telomerase activity appears elevated in most (85–90%) cancer cells.[24,30,154] However, some cells that lack telomerase activity still exhibit a high level of *hTERT* transcription. In these cases, regulation at the level of alternative splicing leads to the skipping of exons that encode reverse transcriptase function; thus, any translation product would not form an active enzyme.[155] In mice, the deletion of either TER or TERT leads to telomere shortening, genomic instability, aneuploidy, telomeric fusion, and aging-related phenotypes.[156,157] Therefore, telomerase dysfunction may result in defects in various highly proliferative cells/tissues, ultimately leading to aging-related degenerative diseases.[158] Corroborating with this, the overexpression of TERT can dramatically increase the life span of mice in the background of the overexpression of tumor suppressor genes, such as p53, p16, and p19, indicating that in mammals, TERT must have an antiaging activity.[159]

3.2. Telomere replication in the absence of telomerase

Although telomerase is considered the main mechanism by which telomeres are maintained, there exist some alternative mechanisms that are activated to maintain telomeres when cells lack telomerase activity. These mechanisms are largely based on recombination events that amplify or rearrange previously existing telomeric sequences,[160,161] and these mechanisms appear to be complementary to both the telomerase method and the method implicated in retrotransposition.[162] These alternative mechanisms were first described in budding yeasts that were able to survive and perform telomere elongation despite the absence of a functional telomerase.[29] It was later verified that this phenomenon was dependent on RAD52, a protein implicated in homologous recombination.[160]

In cancer cells, telomere tracts are also maintained by telomerase in the majority of cases.[154] However, approximately 10–15% of tumor cells elongate their telomeres using one or more alternative mechanisms referred to as ALT.[30,163] Similarly, immortalized cells can also elongate their telomeres using either telomerase[164] or ALT.[165] The ALT mechanism is based on a process that resembles recombination among telomeric DNA sequences.

In this case, one telomere aligns with complementary sequences in another telomere, and telomeric DNA is exchanged.[166] ALT-mediated telomere length maintenance is usually inhibited in normal cells by anti-recombinogenic factors, such as telomeric proteins (e.g., the shelterin components TRF2 and POT1), which have been demonstrated to repress recombination in telomerase-positive murine cells.[167,168]

ALT-positive human cancer cells share many features with normal cells, such as the presence of the telomeric repeat TTAGGG, the single-stranded 3′ G-overhang, the protein complex of shelterin, and several other associated proteins and the formation of T-loops. In contrast, ALT-positive human cancer cells also possess some unique characteristics, which could be used as ALT markers, such as the existence of very long, heterogeneous telomeres and the presence of extrachromosomal telomeric DNA in the form of T-circles (double-stranded telomeric circles), C-circles, or G-circles (partially single-stranded circles rich in C or G nucleotides, respectively), linear dsDNA, large-scale epigenetic modifications at chromosome termini, and high rates of structural and numerical chromosome instability. Among these markers, C-circles appear to be the most specific characteristic able to indicate the presence of ALT and have been suggested as the best ALT marker identified to date. The existence of such a marker would facilitate the development of anticancer drugs targeting the ALT mechanism and would enable the early detection of ALT-positive cancers.[30] An additional remarkable characteristic of ALT-positive cells is the presence of telomeric DNA associated with promyelocytic leukemia nuclear bodies (PML-NBs), also known as ALT-associated PML-NBs,[169] which are composed mainly of proteins such as PML, SP100, SUMO, telomeric chromatin proteins (e.g., TRF1, TRF2, POT1, and RAP1), and repair and recombination factors (e.g., components of the 9–1–1 complex and phosphorylated histone H2A.X, RAD51, and RAD52, among others). Although the molecular function of these telomere-associated PML-NBs remains unknown, their colocalization with recombination proteins supports the theory that homologous recombination may indeed be associated with ALT.[57,170]

There are other telomere-lengthening mechanisms that occur in the absence of telomerase; these mechanisms have been extensively reviewed by others and are only briefly mentioned here. The mosquito *Anopheles gambiae*, the fly *D. melanogaster*, and some species of plants are other examples of organisms that do not use telomerase but, instead, elongate telomeres using recombination.[162,171] *Drosophila*, for instance, lacks telomerase activity and, unlike the majority of organisms, exhibits long tandem arrays composed of three non-LTR retrotransposons, HeT-A, TART, and TAHRE, instead of simple telomeric repeats. These were the first transposable elements shown to play an important role in cell structure.[28,162,172]

In addition, in the protozoa *T. brucei*, an unknown mechanism is able to stabilize critically short telomeres generated by knocking out the *TERT* gene. These short telomeres are transcriptionally inactive, and they shorten progressively without leading to cell senescence due to their stability despite the absence of telomerase.[173–175] The mechanism by which these short telomeres are stabilized has not yet been elucidated, but it is known that the telomerase-deficient strains switch VSG genes by duplicative gene conversion, which occurs more frequently than in wild-type strains, which exhibit longer telomeres. Moreover, it was observed that shorter chromosomes never underwent fusion and that telomere stabilization was sufficient to preserve genomic integrity, with no apparent effects on long-term population growth.[176]

4. TELOMERIC CHROMATIN: IMPLICATIONS FOR END PROTECTION, TELOMERE REPLICATION, AND TELOMERE LENGTH REGULATION

Telomeric chromatin is mainly composed of proteins that interact specifically with double- or single-stranded telomeric DNA or with each other, forming highly ordered and dynamic complexes. These protein–DNA and protein–protein interactions are responsible for maintaining telomere homeostasis. Most of these proteins are characterized by the presence of conserved structural domains and linear motifs, which determine different types of interactions and whether these protein complexes act as positive or negative regulators of telomere length. The balanced relationship between these proteins and telomeres ensures that telomerase will be recruited only when at least one telomere is short and, thus, when cells are in danger, by signaling that chromosome ends are behaving as DNA double-strand breaks. The elongation of the G-strand by telomerase enables telomeres to reach an appropriate length, allowing the subsequent binding of specific end-binding telomeric complexes and the recruitment of the conventional DNA polymerase machinery to perform fill-in synthesis of the C-strand and to prevent any inappropriate repair reactions.[177–179]

In addition, telomere length regulation in most organisms involves the accessibility of the telomerase to telomeres, which may occur at different levels. For example, double-stranded telomere-binding proteins, such as

Rap1 (repressor/activator protein 1) in budding yeast, are involved in telomere length regulation.[180,181] There is evidence to support a "Rap1 protein-counting model," indicating that the presence of a folded chromatin structure at telomeres prevents telomerase access. Telomeric dsDNA-binding proteins, such as TRF1 in humans[182] and Taz1 in fission yeast,[183] may act by a similar mechanism. Another way to make telomeres inaccessible to telomerase is the formation of structures such as T-loops and G-quadruplex[184] (see Section 2.1). Proteins that bind to the 3′ ssDNA tail, in turn, are involved in regulating the access of telomerase to the 3′ end; these proteins usually act as both positive and negative regulators of telomere elongation and end capping.[178,185]

Recently, it was shown that telomeric transcripts known as TERRA[34] or TelRNA[186] also contribute to telomere length maintenance by their negative influence on telomerase activity.

The following text summarizes the most important topics related to the components of telomeric chromatin in model organisms, and it discusses recent findings that reveal how interactions among these components directly affect telomere length homeostasis.

4.1. CST: The major telomere end-binding complex on eukaryotic telomeres

Studies of telomere-associated proteins in a variety of organisms have revealed that, throughout evolution, telomere maintenance mechanisms employ common structural elements.[185] Primary findings about TEBPs, telomeric proteins from the ciliate protozoa *Oxytricha nova*, indicate that TEBPs form both homo- and heterodimeric complexes composed of α and β subunits. TEBPs bind specifically to telomeric single-stranded T4G4 repeats and to each other using OB-fold domains, which share structural similarities with the components of the replication protein A (RPA) heterotrimer.[187,188] TEBP-α binds directly to DNA and shares homology with other end-binding proteins, such as the shelterin component POT1 and the yeast Cdc13, whereas TEBP-β, which does not bind DNA, shows similarities with another shelterin component, TPP1.[189–191] Complexed as α-homodimers, TEBPs function as telomerase recruiters, and as α–β heterodimers, they play the opposite role of blocking telomerase access to the ciliate chromosome ends.[189] Thus, this minimal telomeric complex is able to control telomerase action and, therefore, regulate telomere length.

Telomere end-binding complexes similar to TEBP have also been described in yeast, fungi, plants, and vertebrates. These organisms contain the end-binding complex known as CST, which is also composed of proteins that are considered RPA-like. The CST complex was first discovered in the budding yeast *Saccharomyces cerevisiae* and is composed of three essential proteins: Cdc13, Stn1, and Ten1[192] (Fig. 1.4A). Of these proteins, Stn1 and Ten1 are the most conserved because their orthologs are present in CST-like complexes from a diverse set of species. In all cases, Stn1 and Ten1 share structural similarities with RPA32 and RPA14, respectively. A less-conserved large subunit, which is called Cdc13 in yeast and CTC (conserved telomere maintenance component 1) in other eukaryotes, contains multiple OB-folds with structural homology to the OB-folds from RPA70.[185] Mutations in CTC1 lead to a rare inherited human telomere syndrome, known as Coats plus.[128] In *S. cerevisiae*, ScCST associates with the telomeric 3′ overhang to modulate replication termination and terminal processing and to stimulate priming at the C-strand gap that remains after telomerase elongation.[177] Furthermore, when only Cdc13 is bound to the telomeric 3′ G-overhang, it may mediate the recruitment of telomerase to chromosome ends in late S phase through a specific interaction with the Est1 telomerase subunit.[74,193,194] In this case, *S. cerevisiae* telomerase activity appears to also be regulated by two components of the trimeric RPA complex, which have been found to associate with Est1 during telomere replication.[195] In contrast, human CST (hCST), for example, exhibits slightly distinct functions at telomeres compared with ScCST. hCST works to protect telomeres from eliciting a DNA damage response; to stimulate DNA polymerase α-primase, thus increasing its affinity for the DNA template; and, in a more recent discovery, to specifically bind to telomerase-extended telomeric 3′ overhangs, acting as a terminator of telomerase activity.[21,196] Moreover, hCST specifically binds to telomeric G-rich ssDNA in a size-dependent manner, although it can also bind to long (>50 nt) ssDNA in a sequence-independent manner.[197] Thus, unlike the ScCST complex, ssDNA binding by hCST requires an intact trimeric complex, although the protein components do not possess significant ssDNA-binding activity on their own.[142] This is a huge difference compared with ScCST because yeast Cdc13 binds the telomeric 3′ G-overhang independently of its partners to mediate the recruitment of telomerase.[94]

Thus, a common feature of TEBPs, CST, and CST-like complexes is the presence of OB-fold domains structurally similar to that found in RPA.[198] RPA is an evolutionarily conserved heterotrimeric complex that binds ssDNA in a nonspecific manner[199] and is required for most aspects of DNA metabolism in yeast and some protozoa, including replication, repair,

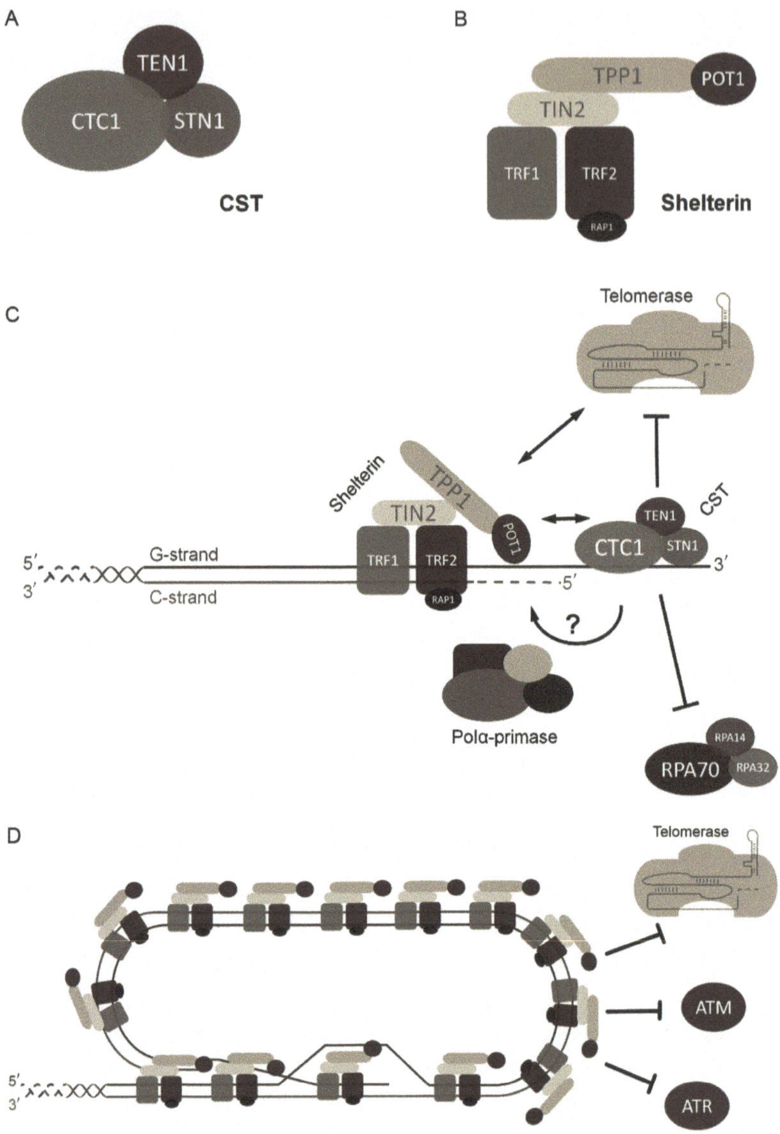

Figure 1.4 *Mammalian telomeric chromatin.* (A) The end-binding CST trimeric complex. (B) The six-member shelterin complex and its specific interactions. (C) The orchestrated actions of shelterin and CST to regulate telomerase access, which promotes telomere elongation and C-strand fill-in synthesis avoiding RPA access. At the telomere ends, telomerase can be recruited by the TPP1/POT1 interaction. The telomere extension by telomerase is terminated when the CST complex binds to the newly synthesized, single-stranded 3′ G-overhang (protruding 3′ single-strand). CST also interacts with DNA polα-primase to promote C-strand fill-in synthesis (dashed line). The CST complex inhibits the binding of RPA to telomeres due to CST higher affinity for single-stranded G-rich DNA. (D) Telomeres in a closed T-loop configuration. Most shelterin components are found interacting with double-stranded telomeric DNA and TPP1/POT1 that are found associated with the 3′ G-overhang and D-loop. T-loop avoids the access of telomerase and the checkpoint kinases ATM and ATR to telomeres.

homologous recombination, and telomere maintenance.[192,195,200–202] Although the protein components of all these complexes share significant structural features with RPA, they appear to perform different biological functions with considerable species-specific variations in the activity of each component. CSTs are actually considered a telomere-specific RPA-like heterotrimer (t-RPA), which contributes to both chromosome end protection and DNA replication at chromosome termini.[178,185]

4.2. Shelterin: A conserved, double-stranded, telomeric protein complex that associates with the telomere end-binding CST complex to maintain telomere homeostasis

Most eukaryotes also possess protein complexes that associate with double-stranded telomeric DNA (dsDNA). Budding yeasts employ a distinct mechanism of telomere maintenance through Rap1 and its interacting factors (Rif1, Rif2, and Sir4).[180,203] In addition to being a transcriptional activator, yeast Rap1 binds telomeric DNA as a monomer via two Myb-like DNA-binding domains.[204] The amount of Rap1 bound to telomeres resembles a "protein-counting" mechanism, which functions as a negative regulator of telomere length.[180,205,206]

In mammals and fission yeasts, for example, six protein complexes termed shelterin (Fig. 1.4B) and shelterin-like, respectively, are formed by the double-stranded telomeric proteins TRF1 and TRF2 (TTAGGG repeat-binding factors 1 and 2) in mammals and Taz1 (telomere length regulator taz1) in fission yeast. In both complexes, these proteins are connected to each other by protein–protein interactions. TFRs, for example, bind directly to DNA as homodimers via independent Myb-like homeodomains.[207,208] TFRs also use specific domains or linear motifs to interact with other proteins, such as the homologue of the yeast Rap1 protein, which does not bind DNA in this case, but interacts with TRF2 and TIN2 (TRF1-interacting protein 1). In turn, TIN2 connects TRF1 to TRF2 and tethers TPP1 (TINT1/PIP1/PTOP1)/POT1 (protection of telomeres 1) to TRF1 and TRF2, contributing to the stabilization of TRF2 on telomeres.[209,210] Thus, TIN2 bridges the double-stranded telomeric complex (TRF1/TRF2–RAP1) to the G-rich single-stranded telomeric complex (TPP1/POT1). The POT1 protein binds to 3′ G-rich overhangs using an OB-fold domain. Similar to other end-binding factors (e.g., Cdc13), POT1 is considered an RPA-like protein due to its structural similarities with RPA70. Unlike RPA, but similar to other TEBPs, POT1 exhibits strong binding specificity for single-stranded G-rich telomeric sequences.[211,212] This single-stranded-binding activity of POT1 may block telomerase from gaining access to the 3′ telomere terminus.[213] In contrast, a POT1 deficiency can lead to uncapped telomeres, which triggers DNA damage checkpoint activation via an ATR (ATM/Rad3-related protein)-dependent DNA damage response pathway, subsequently inducing aberrant homologous recombination at telomeres and formation of telomere circles.[214] This phenomenon explains why shelterin is considered a POT1-dependent protein complex that is crucial for chromosome end protection and telomere length regulation.[177] TRF2 is another pivotal shelterin component because it is directly involved in T-loop formation and stabilization. The depletion of TRF2, unlike that of POT1, leads to telomere deprotection and the formation of dicentric chromosomes because it elicits an ATM (ataxia telangiectasia mutated)-dependent DNA damage response, which results in chromosome end fusion by NHEJ factors.[33]

Nontelomeric factors, such as components of distinct DNA repair machineries, are also associated with telomeric chromatin, where they play roles distinct from that played during the damage response. An example is the KU70/80 heterodimer, which promotes the NHEJ pathway.[215] Interestingly, KU heterodimers are found as an integral component of the terminal telomeric cap in budding yeast via their interaction with Sir4 in addition to associating with the yeast TER component (TLC1).[123,216] The interaction of KU proteins with yeast telomeric chromatin forms a specific architecture, which plays a central role in telomere function and maintenance.[217] The KU heterodimer can also interact with the shelterin components TRF1, TRF2, and RAP1, which favors an indirect interaction with telomeres of distinct eukaryotes without engaging in activities that could pose a threat to telomeres.[41]

Recent findings have contradicted the prevailing view that shelterin and CST represent two distinct telomere-capping complexes that evolved independently. The depletion of TPP1/POT1 increases the hCST–telomere association, suggesting that the two complexes compete for binding to telomeric 3′ overhangs.[196] In addition, the CST–DNA interaction, which is more common during S/G2 phase, limits telomerase action.[21,142] In fact, there is ample evidence that the formation of a trimeric CST complex at the G-rich 3′ overhang leads to telomerase inhibition while, in parallel, mediating a physical interaction with DNA polymerase-α, suggesting that CST coordinates both telomere elongation by telomerase and C-strand fill-in synthesis to complete telomere replication.[21,142] It was recently proposed that CST limits telomerase action on individual telomeres by binding to telomerase-extended telomeres, with the result that binding and extension events occur once per cell cycle.[196] Figure 1.4C shows a diagram summarizing the coordinated actions of shelterin and hCST on human telomeres.

5. TERRA: THE TELOMERIC RNA TRANSCRIPT

It was recently shown that telomeric DNA from a large variety of eukaryotes could be transcribed by RNA polymerase II into long, noncoding TERRA molecules.[218] TERRA originates from the telomeric C-rich strand and gives rise to UUAGGG repeat-containing telomeric transcripts that range in size from 100 bp to 9 kb. In mammals and yeasts, TERRA can be found as either 3′ end-polyadenylated, poly(A)(+), or poly(A)(−) RNAs. TERRA transcription is regulated by telomere-binding proteins in a chromosome end-specific manner in yeast[219]; in humans, it is driven by CpG island promoters located at subtelomeres.[220] The molecular mechanisms and factors that control TERRA expression levels are being studied intensively. TERRA partners and regulatory factors include members of the heterogeneous nuclear RNP family (e.g., hnRNPA1), heterochromatin proteins (e.g., heterochromatin protein 1 (HP1) and histone H3 trimethyl K9), and telomeric proteins from the shelterin complex (e.g., TRF2). The abundance of TERRA is highly dependent on cell cycle stage, developmental status, telomere length, cellular stress, tumor stage, and chromatin structure,[218] possibly reflecting distinct biological roles of TERRA–protein complexes. For example, contrary to expectations, the transcription of TERRA does not inhibit telomerase[221]; instead, it coordinates the recruitment of telomerase molecules to the shortest telomeres from which the transcript originated. TERRA appears to be directly involved in the nucleation of telomerase molecules into clusters prior to their recruitment to short telomeres.[35] The interaction of TERRA with human telomerase *in vitro* has been shown to occur by base pairing with the template sequence of TER and by contact with the TERT subunit, leading to the inhibition of telomerase activity.[34] In budding yeast, the induction of TERRA leads to telomere shortening, although this involves the activation of the chromosome end trimming exonuclease 1 at chromosome ends rather than the inhibition of telomerase.[222] Redon *et al.*[223] proposed a three-state model of telomerase regulation by TERRA, whereby an excess of TERRA after S phase may sequester hnRNPA1 from telomeric DNA (hnRNPA1 was shown to bind single-stranded telomeric DNA without interfering with telomerase activity[224]). Consequently, the telomeric 3′ overhang is freed from RPA, which promotes binding by POT1–TPP1[225] and, subsequently, the recruitment of telomerase. Thus, *in vitro* telomerase activity is maintained as long as the levels of TERRA and hnRNPA1 remain balanced.

Once telomeres are elongated, the transcription of TERRA is inhibited by heterochromatin protein HP1α and by the increased methylation of histone H3K9, which becomes trimethylated and dense at telomeres. In conclusion, TERRA represses its own expression by negative feedback and upon telomere elongation.[226]

6. CONSEQUENCES OF TELOMERE DEPROTECTION

Telomeres cap chromosome ends while avoiding nucleolytic attack, and they distinguish chromosome termini from DNA double-strand breaks. Telomeres naturally shorten and become dysfunctional during aging in human somatic cells because most of these cells lack telomerase activity.[43] When telomeres become critically short, a cellular response is triggered, signaling cells to exit the cell cycle and to senesce (Fig. 1.5A). This process indicates that the cells have reached their maximum proliferation capacity, known as the Hayflick limit.[227] The rare cells that escape this natural crisis have already lost critical cell cycle checkpoints and components of the DNA damage repair machinery; they emerge with multiple chromosomal abnormalities and mutations, as well as with an active mechanism of telomere maintenance (the reactivation of telomerase or ALT). A similar phenotype has been described in yeast mutants with a telomerase deletion.[228] Thus, the cells that bypassed the crisis are now considered immortalized. Some of these cells also bear mutations in genes responsible for inducing tumorigenesis, making them prone to promote cancer formation.[78]

Dysfunctional telomeres can also arise through other mechanisms, such as mutations in telomerase components (TERT, TER, and accessory proteins) or the genomic deletion of a telomere-binding protein.

Deficiencies in components of the telomerase holoenzyme are implicated in several rare genetically inherited disorders, such as aplastic anemia, dyskeratosis congenita, and idiopathic pulmonary fibrosis.[128,229] Dyskeratosis congenita, for example, is a disease associated with mutations in the dyskerin gene (*DKC1*), which encodes dyskerin, a telomerase holoenzyme component and a nucleolar protein involved in small RNA biogenesis. Patients with dyskeratosis exhibit short telomeres and die at an early age, presenting with symptoms that vary from abnormal skin pigmentation, nail dystrophy, and leukoplakia to bone marrow failure and

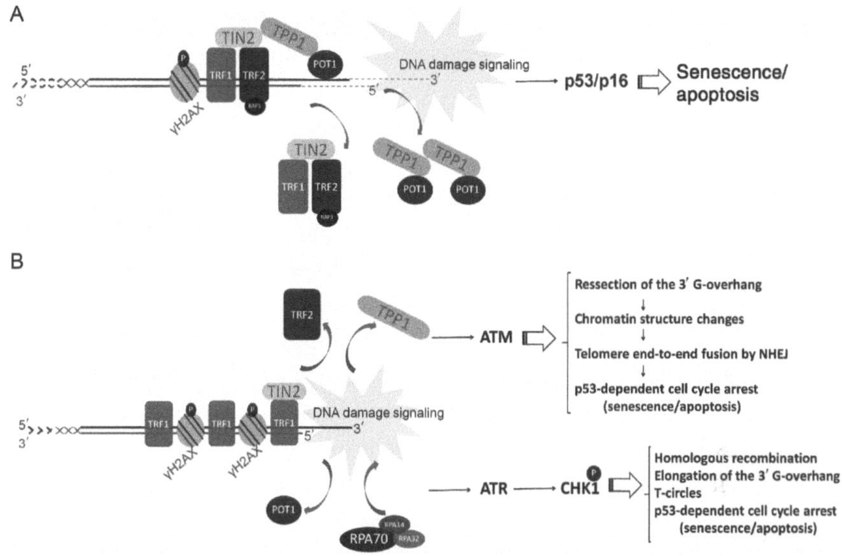

Figure 1.5 *The many faces of telomere deprotection.* (A) Short telomeres can arise naturally with age in cells with low proliferative capacity that does not possess telomerase activity. The same can occur in cells that lack telomerase activity due to mutations or deletions of components of the telomerase holoenzyme. In this case, if an active telomere replication mechanism (telomerase or ALT) is not activated, the telomeres will become critically short, which triggers a cellular response, signaling the cell to exit the cell cycle, senesce, and die. (B) Uncapped telomeres generated by mutations or deletions in a telomeric protein. Long telomeres lacking TRF2 or POT1 trigger ATM- or ATR-dependent DNA damage responses, respectively, leading to telomere end-to-end fusions by NHEJ (depletion of TRF2 and TPP1) or homologous recombination between telomeres (loss of POT1). In both (A) and (B), a p53-dependent cell cycle arrest induces cells to senesce or dye.

pulmonary fibrosis.[230] Details about dyskeratosis congenita will be presented in chapter 6 in this book.

The importance of a proper cap structure is highlighted by the finding that dysfunctional telomeres formed by mutating components of telomeric chromatin can recombine or fuse, leading to major genomic rearrangements. As described above, the loss of the shelterin component POT1 or TRF2 leads to the activation of the ATR and ATM checkpoint kinases, respectively, at telomeres, which elicit distinct DNA damage responses.[231] The loss of POT1 also leads to the aberrant accumulation of RPA at telomeres, which induces an ATR-dependent response that leads to the phosphorylation of CHK1 (checkpoint kinase 1).[232] In addition, POT1-deficient mouse cells elongate telomeres and show aberrant homologous recombination, including telomere sister chromatid exchange and formation of T-circles, resulting in a p53-dependent cell cycle arrest indistinguishable from replicative senescence[214] (Fig. 1.5B). POT1 loss alone does not induce telomere fusion by NHEJ, only in the absence of a functional TRF2. Curiously, both POT1 and TPP1 are unable to efficiently prevent RPA binding to telomeric ssDNA *in vitro*, suggesting that these proteins antagonize RPA binding.[233] Actually, TERRA, hnRNPA1, and POT1 act in concert *in vivo* to displace RPA from the G-rich telomeric strand after telomere replication, thus promoting telomere capping.[225] In addition, the knockdown of the POT1 partner TPP1 elicits a p53-dependent growth arrest and an ATM-dependent DNA damage response at telomeres. In contrast, in the absence of an ATM-dependent DNA damage response, TPP1 depletion causes chromosome instability and tumorigenesis in mice.[231]

Telomeres can also lose their protective function despite the presence of long stretches of telomeric repeats, a well-known phenomenon called "telomere uncapping." A good example of this phenomenon occurs after the removal of TRF2 from telomeres either using a dominant-negative allele or in TRF2-knock-out mice. In both cases, after a few cell divisions, all the chromosome ends become fused because they were identified as DNA double-strand breaks elicited by an ATM-dependent response. These and other types of dysfunctional telomeres can be observed *in vitro* as cytogenetic figures termed TIFs (telomere dysfunction-induced foci). TIFs are also detected in cells undergoing replicative senescence and at sites of breakage and repair following damage induced by irradiation. At dysfunctional telomeres, the telomeric chromatin is mainly occupied by telomeric proteins (e.g., TRF2), 53BP1 (p53-binding protein 1), γH2AX, ATM, Mre11 (meiotic recombination 11 homologue A), and other proteins involved in the DNA damage response.[9,234,235] Thus, although uncapped telomeres are not broken ends, they are subject to repair by almost the same pathways. For example, restoring telomerase activity can directly elongate a short telomere to a size that will permit capping.[236] Homologous recombination between two short telomeres or one short and one long telomere can also recreate a safe length.[160,237] Both events work very efficiently in yeast telomerase mutants,[238,239] but they are not feasible when all the telomeres are critically short due to a defect in a telomeric protein. In this case, it is more common to observe telomere fusion caused by the action of NHEJ machinery proteins, which induce fusions between two telomeres of different chromosome arms or between identical replicated sister chromatids.[240,241] These differences among types of telomere defect resolution are explained by the finding that the telomere dysfunction generated when cells divide repeatedly in the absence of telomerase affects several or even individual telomeres, but not all telomeres simultaneously. In this case, cells can exhibit some chromosomes with very short telomeres and others with long telomeres. In contrast, the dysfunction elicited by the mutation or deletion of a telomeric protein simultaneously affects all the telomeres and can also induce a variety of phenotypes.[242]

The primary function of telomeres is to prevent chromosome ends from being recognized as broken ends. However, as mentioned above, for a variety of reasons, including telomere uncapping and inherited telomere/telomerase abnormalities, telomere dysfunction can lead to gross chromosomal rearrangements that impair genome stability and, consequently, cellular homeostasis. Controlling the mechanisms underlying telomere maintenance is the main challenge for researchers studying telomeres.

ACKNOWLEDGMENTS

We apologize to the researchers whose work could not be mentioned due to space constraints. This work was supported by FAPESP Grant 2012/50263-5. M. S. S., M. S., and V. S. N. received doctoral and postdoctoral fellowships, respectively, from FAPESP. M. I. N. C. is a CNPq research fellow.

REFERENCES

1. Blackburn EH, Gall JG. A tandemly repeated sequence at the termini of the extrachromosomal ribosomal RNA genes in Tetrahymena. *J Mol Biol.* 1978;120:33–53.
2. Meyne J, Ratliff RL, Moyzis RK. Conservation of the human telomere sequence (TTAGGG)n among vertebrates. *Proc Natl Acad Sci USA.* 1989;86:7049–7053.
3. Pryde FE, Gorham HC, Louis EJ. Chromosome ends: all the same under their caps. *Curr Opin Genet Dev.* 1997;7:822–828.
4. Fajkus J, Sykorova E, Leitch AR. Telomeres in evolution and evolution of telomeres. *Chromosome Res.* 2005;13:469–479.
5. Greider CW, Blackburn EH. Identification of a specific telomere terminal transferase activity in Tetrahymena extracts. *Cell.* 1985;43:405–413.
6. Harley CB, Futcher AB, Greider CW. Telomeres shorten during ageing of human fibroblasts. *Nature.* 1990;345:458–460.
7. Sandell LL, Zakian VA. Loss of a yeast telomere: arrest, recovery, and chromosome loss. *Cell.* 1993;75:729–739.
8. de Lange T, Lundblad V, Blackburn EH. *Telomeres.* Cold Spring Harbor, New York: Cold Spring Harbor Laboratory Press; 2006.
9. O'Sullivan RJ, Karlseder J. Telomeres: protecting chromosomes against genome instability. *Nat Rev Mol Cell Biol.* 2010;11:171–181.
10. Greider CW. Telomeres and senescence: the history, the experiment, the future. *Curr Biol.* 1998;8:R178–R181.
11. Stewart JA, Chaiken MF, Wang F, Price CM. Maintaining the end: roles of telomere proteins in end-protection, telomere replication and length regulation. *Mutat Res.* 2012;730:12–19.

12. Celli GB, de Lange T. DNA processing is not required for ATM-mediated telomere damage response after TRF2 deletion. *Nat Cell Biol*. 2005;7:712–718.
13. Verdun RE, Karlseder J. Replication and protection of telomeres. *Nature*. 2007;447:924–931.
14. Blackburn EH. Telomeres and their synthesis. *Science*. 1990;249:489–490.
15. Greider CW. Telomere length regulation. *Annu Rev Biochem*. 1996;65:337–365.
16. Gilson E, Geli V. How telomeres are replicated. *Nat Rev Mol Cell Biol*. 2007;8:825–838.
17. Londono-Vallejo JA, Wellinger RJ. Telomeres and telomerase dance to the rhythm of the cell cycle. *Trends Biochem Sci*. 2012;37:391–399.
18. Watson JD. Origin of concatemeric T7 DNA. *Nat New Biol*. 1972;239:197–201.
19. Greider CW, Blackburn EH. A telomeric sequence in the RNA of Tetrahymena telomerase required for telomere repeat synthesis. *Nature*. 1989;337:331–337.
20. Olovnikov AM. A theory of marginotomy. The incomplete copying of template margin in enzymic synthesis of polynucleotides and biological significance of the phenomenon. *J Theor Biol*. 1973;41:181–190.
21. Wu P, Takai H, de Lange T. Telomeric 3' overhangs derive from resection by Exo1 and Apollo and fill-in by POT1b-associated CST. *Cell*. 2012;150:39–52.
22. Stewart JA, Wang F, Chaiken MF, et al. Human CST promotes telomere duplex replication and general replication restart after fork stalling. *EMBO J*. 2012;31:3537–3549.
23. Autexier C, Lue NF. The structure and function of telomerase reverse transcriptase. *Annu Rev Biochem*. 2006;75:493–517.
24. Kim NW, Piatyszek MA, Prowse KR, et al. Specific association of human telomerase activity with immortal cells and cancer. *Science*. 1994;266:2011–2015.
25. Cao Y, Li H, Deb S, Liu JP. TERT regulates cell survival independent of telomerase enzymatic activity. *Oncogene*. 2002;21:3130–3138.
26. Smith LL, Coller HA, Roberts JM. Telomerase modulates expression of growth-controlling genes and enhances cell proliferation. *Nat Cell Biol*. 2003;5:474–479.
27. Rahman R, Latonen L, Wiman KG. hTERT antagonizes p53-induced apoptosis independently of telomerase activity. *Oncogene*. 2005;24:1320–1327.
28. Pardue ML, DeBaryshe PG. Drosophila telomeres: a variation on the telomerase theme. *Fly (Austin)*. 2008;2:101–110.
29. Lundblad V, Blackburn EH. An alternative pathway for yeast telomere maintenance rescues est1- senescence. *Cell*. 1993;73:347–360.
30. Cesare AJ, Reddel RR. Alternative lengthening of telomeres: models, mechanisms and implications. *Nat Rev Genet*. 2010;11:319–330.
31. Chen Q, Ijpma A, Greider CW. Two survivor pathways that allow growth in the absence of telomerase are generated by distinct telomere recombination events. *Mol Cell Biol*. 2001;21:1819–1827.
32. Stevenson JB, Gottschling DE. Telomeric chromatin modulates replication timing near chromosome ends. *Genes Dev*. 1999;13:146–151.
33. Diotti R, Loayza D. Shelterin complex and associated factors at human telomeres. *Nucleus*. 2012;2:119–135.
34. Azzalin CM, Reichenbach P, Khoriauli L, Giulotto E, Lingner J. Telomeric repeat containing RNA and RNA surveillance factors at mammalian chromosome ends. *Science*. 2007;318:798–801.
35. Cusanelli E, Romero CA, Chartrand P. Telomeric noncoding RNA TERRA is induced by telomere shortening to nucleate telomerase molecules at short telomeres. *Mol Cell*. 2013;51:780–791.
36. Cesare AJ, Hayashi MT, Crabbe L, Karlseder J. The telomere deprotection response is functionally distinct from the genomic DNA damage response. *Mol Cell*. 2013;51:141–155.

37. Jacobs JJ. Loss of telomere protection: consequences and opportunities. *Front Oncol.* 2013;3:88.
38. Wellinger RJ, Sen D. The DNA structures at the ends of eukaryotic chromosomes. *Eur J Cancer.* 1997;33:735–749.
39. Shampay J, Szostak JW, Blackburn EH. DNA sequences of telomeres maintained in yeast. *Nature.* 1984;310:154–157.
40. Henderson ER, Blackburn EH. An overhanging 3′ terminus is a conserved feature of telomeres. *Mol Cell Biol.* 1989;9:345–348.
41. Nandakumar J, Cech TR. Finding the end: recruitment of telomerase to telomeres. *Nat Rev Mol Cell Biol.* 2013;14:69–82.
42. Cano MI. Telomere biology of trypanosomatids: more questions than answers. *Trends Parasitol.* 2001;17:425–429.
43. Aubert G, Lansdorp PM. Telomeres and aging. *Physiol Rev.* 2008;88:557–579.
44. Cohn M. Molecular diversity of telomeric sequences. In: Nosek J, Tomaska L, eds. *Origin and Evolution of Telomeres.* Austin, TX: Landes, Bioscience; 2008:70–77.
45. Casacuberta E, Pardue ML. HeT-A and TART, two Drosophila retrotransposons with a bona fide role in chromosome structure for more than 60 million years. *Cytogenet Genome Res.* 2005;110:152–159.
46. Melek M, Davis BT, Shippen DE. Oligonucleotides complementary to the Oxytricha nova telomerase RNA delineate the template domain and uncover a novel mode of primer utilization. *Mol Cell Biol.* 1994;14:7827–7838.
47. McCormick-Graham M, Romero DP. A single telomerase RNA is sufficient for the synthesis of variable telomeric DNA repeats in ciliates of the genus Paramecium. *Mol Cell Biol.* 1996;16:1871–1879.
48. Cohn M, Blackburn EH. Telomerase in yeast. *Science.* 1995;269:396–400.
49. Fulton TB, Blackburn EH. Identification of Kluyveromyces lactis telomerase: discontinuous synthesis along the 30-nucleotide-long templating domain. *Mol Cell Biol.* 1998;18:4961–4970.
50. Kipling D, Cooke HJ. Hypervariable ultra-long telomeres in mice. *Nature.* 1990;347:400–402.
51. Witzany G. The viral origins of telomeres and telomerases and their important role in eukaryogenesis and genome maintenance. *Biosemiotics.* 2008;1:16.
52. Nosek J, Kosa P, Tomaska L. On the origin of telomeres: a glimpse at the pre-telomerase world. *Bioessays.* 2006;28:182–190.
53. Griffith JD, Comeau L, Rosenfield S, et al. Mammalian telomeres end in a large duplex loop. *Cell.* 1999;97:503–514.
54. de Lange T. T-loops and the origin of telomeres. *Nat Rev Mol Cell Biol.* 2004;5:323–329.
55. Tomaska L, Willcox S, Slezakova J, Nosek J, Griffith JD. Taz1 binding to a fission yeast model telomere: formation of telomeric loops and higher order structures. *J Biol Chem.* 2004;279:50764–50772.
56. Stansel RM, de Lange T, Griffith JD. T-loop assembly in vitro involves binding of TRF2 near the 3′ telomeric overhang. *EMBO J.* 2001;20:5532–5540.
57. Nabetani A, Ishikawa F. Alternative lengthening of telomeres pathway: recombination-mediated telomere maintenance mechanism in human cells. *J Biochem.* 2011;149:5–14.
58. Pedroso IM, Hayward W, Fletcher TM. The effect of the TRF2 N-terminal and TRFH regions on telomeric G-quadruplex structures. *Nucleic Acids Res.* 2009;37:1541–1554.
59. Wang H, Nora GJ, Ghodke H, Opresko PL. Single molecule studies of physiologically relevant telomeric tails reveal POT1 mechanism for promoting G-quadruplex unfolding. *J Biol Chem.* 2011;286:7479–7489.

60. Yanez GH, Khan SJ, Locovei AM, Pedroso IM, Fletcher TM. DNA structure-dependent recruitment of telomeric proteins to single-stranded/double-stranded DNA junctions. *Biochem Biophys Res Commun*. 2005;328:49–56.
61. Oganesian L, Bryan TM. Physiological relevance of telomeric G-quadruplex formation: a potential drug target. *Bioessays*. 2007;29:155–165.
62. Wang Q, Liu JQ, Chen Z, et al. G-quadruplex formation at the 3′ end of telomere DNA inhibits its extension by telomerase, polymerase and unwinding by helicase. *Nucleic Acids Res*. 2011;39:6229–6237.
63. McClintock B. The stability of broken ends of chromosomes in Zea Mays. *Genetics*. 1941;26:234–282.
64. Muller HJ. The remaking of chromosomes. *Collecting Net*. 1938;13:15.
65. Blackburn EH, Challoner PB. Identification of a telomeric DNA sequence in Trypanosoma brucei. *Cell*. 1984;36:447–457.
66. Cohn M, McEachern MJ, Blackburn EH. Telomeric sequence diversity within the genus Saccharomyces. *Curr Genet*. 1998;33:83–91.
67. Szostak JW, Blackburn EH. Cloning yeast telomeres on linear plasmid vectors. *Cell*. 1982;29:245–255.
68. Bernards A, Michels PA, Lincke CR, Borst P. Growth of chromosome ends in multiplying trypanosomes. *Nature*. 1983;303:592–597.
69. Blackburn EH, Greider CW, Henderson E, Lee MS, Shampay J, Shippen-Lentz D. Recognition and elongation of telomeres by telomerase. *Genome*. 1989;31:553–560.
70. Lingner J, Cech TR. Purification of telomerase from Euplotes aediculatus: requirement of a primer 3' overhang. *Proc Natl Acad Sci USA*. 1996;93:10712–10717.
71. Nakamura TM, Morin GB, Chapman KB, et al. Telomerase catalytic subunit homologs from fission yeast and human. *Science*. 1997;277:955–959.
72. Harrington L. Biochemical aspects of telomerase function. *Cancer Lett*. 2003;194:139–154.
73. Cohen SB, Graham ME, Lovrecz GO, Bache N, Robinson PJ, Reddel RR. Protein composition of catalytically active human telomerase from immortal cells. *Science*. 2007;315:1850–1853.
74. Smogorzewska A, de Lange T. Regulation of telomerase by telomeric proteins. *Annu Rev Biochem*. 2004;73:177–208.
75. Chen JL, Greider CW. An emerging consensus for telomerase RNA structure. *Proc Natl Acad Sci USA*. 2004;101:14683–14684.
76. Harrington L. Making the most of a little: dosage effects in eukaryotic telomere length maintenance. *Chromosome Res*. 2005;13:493–504.
77. Lingner J, Hughes TR, Shevchenko A, Mann M, Lundblad V, Cech TR. Reverse transcriptase motifs in the catalytic subunit of telomerase. *Science*. 1997;276:561–567.
78. Meyerson M, Counter CM, Eaton EN, et al. hEST2, the putative human telomerase catalytic subunit gene, is up-regulated in tumor cells and during immortalization. *Cell*. 1997;90:785–795.
79. Giardini MA, Lira CB, Conte FF, et al. The putative telomerase reverse transcriptase component of Leishmania amazonensis: gene cloning and characterization. *Parasitol Res*. 2006;98:447–454.
80. Cano MI, Dungan JM, Agabian N, Blackburn EH. Telomerase in kinetoplastid parasitic protozoa. *Proc Natl Acad Sci USA*. 1999;96:3616–3621.
81. Lue NF. A physical and functional constituent of telomerase anchor site. *J Biol Chem*. 2005;280:26586–26591.
82. Moriarty TJ, Ward RJ, Taboski MA, Autexier C. An anchor site-type defect in human telomerase that disrupts telomere length maintenance and cellular immortalization. *Mol Biol Cell*. 2005;16:3152–3161.

83. Romi E, Baran N, Gantman M, et al. High-resolution physical and functional mapping of the template adjacent DNA binding site in catalytically active telomerase. *Proc Natl Acad Sci USA*. 2007;104:8791–8796.
84. Wyatt HD, Tsang AR, Lobb DA, Beattie TL. Human telomerase reverse transcriptase (hTERT) Q169 is essential for telomerase function in vitro and in vivo. *PLoS One*. 2009;4:e7176.
85. Jurczyluk J, Nouwens AS, Holien JK, et al. Direct involvement of the TEN domain at the active site of human telomerase. *Nucleic Acids Res*. 2011;39:1774–1788.
86. Armbruster BN, Banik SS, Guo C, Smith AC, Counter CM. N-terminal domains of the human telomerase catalytic subunit required for enzyme activity in vivo. *Mol Cell Biol*. 2001;21:7775–7786.
87. Talley JM, DeZwaan DC, Maness LD, Freeman BC, Friedman KL. Stimulation of yeast telomerase activity by the ever shorter telomere 3 (Est3) subunit is dependent on direct interaction with the catalytic protein Est2. *J Biol Chem*. 2011;286:26431–26439.
88. Gillis AJ, Schuller AP, Skordalakes E. Structure of the Tribolium castaneum telomerase catalytic subunit TERT. *Nature*. 2008;455:633–637.
89. Mitchell M, Gillis A, Futahashi M, Fujiwara H, Skordalakes E. Structural basis for telomerase catalytic subunit TERT binding to RNA template and telomeric DNA. *Nat Struct Mol Biol*. 2010;17:513–518.
90. Bryan TM, Goodrich KJ, Cech TR. Telomerase RNA bound by protein motifs specific to telomerase reverse transcriptase. *Mol Cell*. 2000;6:493–499.
91. Cech TR. Beginning to understand the end of the chromosome. *Cell*. 2004;116:273–279.
92. Collins K. The biogenesis and regulation of telomerase holoenzymes. *Nat Rev Mol Cell Biol*. 2006;7:484–494.
93. Lundblad V, Szostak JW. A mutant with a defect in telomere elongation leads to senescence in yeast. *Cell*. 1989;57:633–643.
94. Pennock E, Buckley K, Lundblad V. Cdc13 delivers separate complexes to the telomere for end protection and replication. *Cell*. 2001;104:387–396.
95. Taggart AK, Teng SC, Zakian VA. Est1p as a cell cycle-regulated activator of telomere-bound telomerase. *Science*. 2002;297:1023–1026.
96. Reichenbach P, Hoss M, Azzalin CM, Nabholz M, Bucher P, Lingner J. A human homolog of yeast Est1 associates with telomerase and uncaps chromosome ends when overexpressed. *Curr Biol*. 2003;13:568–574.
97. Snow BE, Erdmann N, Cruickshank J, et al. Functional conservation of the telomerase protein Est1p in humans. *Curr Biol*. 2003;13:698–704.
98. Venteicher AS, Meng Z, Mason PJ, Veenstra TD, Artandi SE. Identification of ATPases pontin and reptin as telomerase components essential for holoenzyme assembly. *Cell*. 2008;132:945–957.
99. Khusial P, Plaag R, Zieve GW. LSm proteins form heptameric rings that bind to RNA via repeating motifs. *Trends Biochem Sci*. 2005;30:522–528.
100. Meier UT. The many facets of H/ACA ribonucleoproteins. *Chromosoma*. 2005;114:1–14.
101. Holt SE, Aisner DL, Baur J, et al. Functional requirement of p23 and Hsp90 in telomerase complexes. *Genes Dev*. 1999;13:817–826.
102. Forsythe HL, Jarvis JL, Turner JW, Elmore LW, Holt SE. Stable association of hsp90 and p23, but Not hsp70, with active human telomerase. *J Biol Chem*. 2001;276:15571–15574.
103. Zappulla DC, Cech TR. Yeast telomerase RNA: a flexible scaffold for protein subunits. *Proc Natl Acad Sci USA*. 2004;101:10024–10029.

104. Stone MD, Mihalusova M, O'Connor CM, Prathapam R, Collins K, Zhuang X. Stepwise protein-mediated RNA folding directs assembly of telomerase ribonucleoprotein. *Nature*. 2007;446:458–461.
105. Hardy CD, Schultz CS, Collins K. Requirements for the dGTP-dependent repeat addition processivity of recombinant Tetrahymena telomerase. *J Biol Chem*. 2001;276:4863–4871.
106. Chen JL, Greider CW. Determinants in mammalian telomerase RNA that mediate enzyme processivity and cross-species incompatibility. *EMBO J*. 2003;22:304–314.
107. Lai CK, Miller MC, Collins K. Roles for RNA in telomerase nucleotide and repeat addition processivity. *Mol Cell*. 2003;11:1673–1683.
108. Chen JL, Blasco MA, Greider CW. Secondary structure of vertebrate telomerase RNA. *Cell*. 2000;100:503–514.
109. Brault ME, D'Souza Y, Autexier C. Telomerase: evolution, structure and function. In: Nosek J, Tomaska L, eds. *Origin and Evolution of Telomeres*. Landes: Bioscience; 2008:1–17.
110. Blackburn EH, Collins K. Telomerase: an RNP enzyme synthesizes DNA. *Cold Spring Harb Perspect Biol*. 2011;3:a003558.
111. Sandhu R, Sanford S, Basu S, et al. A trans-spliced telomerase RNA dictates telomere synthesis in Trypanosoma brucei. *Cell Res*. 2013;23:537–551.
112. Gupta SK, Kolet L, Doniger T, et al. The Trypanosoma brucei telomerase RNA (TER) homologue binds core proteins of the C/D snoRNA family. *FEBS Lett*. 2013;587:1399–1404.
113. Zappulla DC, Goodrich K, Cech TR. A miniature yeast telomerase RNA functions in vivo and reconstitutes activity in vitro. *Nat Struct Mol Biol*. 2005;12:1072–1077.
114. Mitchell JR, Cheng J, Collins K. A box H/ACA small nucleolar RNA-like domain at the human telomerase RNA 3' end. *Mol Cell Biol*. 1999;19:567–576.
115. Lin J, Ly H, Hussain A, et al. A universal telomerase RNA core structure includes structured motifs required for binding the telomerase reverse transcriptase protein. *Proc Natl Acad Sci USA*. 2004;101:14713–14718.
116. Chen JL, Opperman KK, Greider CW. A critical stem-loop structure in the CR4-CR5 domain of mammalian telomerase RNA. *Nucleic Acids Res*. 2002;30:592–597.
117. Brown Y, Abraham M, Pearl S, Kabaha MM, Elboher E, Tzfati Y. A critical three-way junction is conserved in budding yeast and vertebrate telomerase RNAs. *Nucleic Acids Res*. 2007;35:6280–6289.
118. Robart AR, O'Connor CM, Collins K. Ciliate telomerase RNA loop IV nucleotides promote hierarchical RNP assembly and holoenzyme stability. *RNA*. 2010;16:563–571.
119. Tesmer VM, Ford LP, Holt SE, et al. Two inactive fragments of the integral RNA cooperate to assemble active telomerase with the human protein catalytic subunit (hTERT) in vitro. *Mol Cell Biol*. 1999;19:6207–6216.
120. Mitchell JR, Collins K. Human telomerase activation requires two independent interactions between telomerase RNA and telomerase reverse transcriptase. *Mol Cell*. 2000;6:361–371.
121. Egan ED, Collins K. Specificity and stoichiometry of subunit interactions in the human telomerase holoenzyme assembled in vivo. *Mol Cell Biol*. 2010;30:2775–2786.
122. Stellwagen AE, Haimberger ZW, Veatch JR, Gottschling DE. Ku interacts with telomerase RNA to promote telomere addition at native and broken chromosome ends. *Genes Dev*. 2003;17:2384–2395.
123. Gallardo F, Olivier C, Dandjinou AT, Wellinger RJ, Chartrand P. TLC1 RNA nucleo-cytoplasmic trafficking links telomerase biogenesis to its recruitment to telomeres. *EMBO J*. 2008;27:748–757.

124. Egan ED, Collins K. Biogenesis of telomerase ribonucleoproteins. *RNA*. 2012;18:1747–1759.
125. Vulliamy T, Marrone A, Goldman F, et al. The RNA component of telomerase is mutated in autosomal dominant dyskeratosis congenita. *Nature*. 2001;413:432–435.
126. Theimer CA, Finger LD, Trantirek L, Feigon J. Mutations linked to dyskeratosis congenita cause changes in the structural equilibrium in telomerase RNA. *Proc Natl Acad Sci USA*. 2003;100:449–454.
127. Armanios M. Syndromes of telomere shortening. *Annu Rev Genomics Hum Genet*. 2009;10:45–61.
128. Armanios M, Blackburn EH. The telomere syndromes. *Nat Rev Genet*. 2012;13:12.
129. Blasco MA, Lee HW, Hande MP, et al. Telomere shortening and tumor formation by mouse cells lacking telomerase RNA. *Cell*. 1997;91:25–34.
130. Lee HW, Blasco MA, Gottlieb GJ, Horner 2nd JW, Greider CW, DePinho RA. Essential role of mouse telomerase in highly proliferative organs. *Nature*. 1998;392:569–574.
131. Greenberg RA, Chin L, Femino A, et al. Short dysfunctional telomeres impair tumorigenesis in the INK4a(delta2/3) cancer-prone mouse. *Cell*. 1999;97:515–525.
132. Blasco MA. Mice with bad ends: mouse models for the study of telomeres and telomerase in cancer and aging. *EMBO J*. 2005;24:1095–1103.
133. Hemann MT, Strong MA, Hao LY, Greider CW. The shortest telomere, not average telomere length, is critical for cell viability and chromosome stability. *Cell*. 2001;107:67–77.
134. Samper E, Flores JM, Blasco MA. Restoration of telomerase activity rescues chromosomal instability and premature aging in Terc-/- mice with short telomeres. *EMBO Rep*. 2001;2:800–807.
135. Bonetti D, Martina M, Falcettoni M, Longhese MP. Telomere-end processing: mechanisms and regulation. *Chromosoma*. 2013;123:57–66.
136. Greider CW. Telomerase is processive. *Mol Cell Biol*. 1991;11:4572–4580.
137. Lue NF. Adding to the ends: what makes telomerase processive and how important is it? *Bioessays*. 2004;26:955–962.
138. Wallweber G, Gryaznov S, Pongracz K, Pruzan R. Interaction of human telomerase with its primer substrate. *Biochemistry*. 2003;42:589–600.
139. Steitz TA. DNA polymerases: structural diversity and common mechanisms. *J Biol Chem*. 1999;274:17395–17398.
140. Chang M, Arneric M, Lingner J. Telomerase repeat addition processivity is increased at critically short telomeres in a Tel1-dependent manner in Saccharomyces cerevisiae. *Genes Dev*. 2007;21:2485–2494.
141. Wang J, Yu L, Li J, Deng R, Wang X. Characterization of a human telomerase reverse transcriptase sequence containing two antigenic epitopes with high affinity for human leucocyte antigen. *Biotechnol Appl Biochem*. 2007;48:93–99.
142. Chen LY, Lingner J. CST for the grand finale of telomere replication. *Nucleus*. 2013;4:277–282.
143. Teixeira MT, Arneric M, Sperisen P, Lingner J. Telomere length homeostasis is achieved via a switch between telomerase-extendible and -nonextendible states. *Cell*. 2004;117:323–335.
144. Schaetzlein S, Rudolph KL. Telomere length regulation during cloning, embryogenesis and ageing. *Reprod Fertil Dev*. 2005;17:85–96.
145. Djojosubroto MW, Choi YS, Lee HW, Rudolph KL. Telomeres and telomerase in aging, regeneration and cancer. *Mol Cells*. 2003;15:164–175.
146. Sharpless NE, DePinho RA. Telomeres, stem cells, senescence, and cancer. *J Clin Invest* 2004;113:160–168.
147. Shay JW, Wright WE. Senescence and immortalization: role of telomeres and telomerase. *Carcinogenesis*. 2005;26:867–874.

148. Blackburn EH. Cell biology: Shaggy mouse tales. *Nature*. 2005;436:922–923.
149. Xu D, Popov N, Hou M, et al. Switch from Myc/Max to Mad1/Max binding and decrease in histone acetylation at the telomerase reverse transcriptase promoter during differentiation of HL60 cells. *Proc Natl Acad Sci USA*. 2001;98:3826–3831.
150. Lou F, Chen X, Jalink M, et al. The opposing effect of hypoxia-inducible factor-2alpha on expression of telomerase reverse transcriptase. *Mol Cancer Res*. 2007;5:793–800.
151. Hou M, Wang X, Popov N, et al. The histone deacetylase inhibitor trichostatin A derepresses the telomerase reverse transcriptase (hTERT) gene in human cells. *Exp Cell Res*. 2002;274:25–34.
152. Gladych M, Wojtyla A, Rubis B. Human telomerase expression regulation. *Biochem Cell Biol*. 2011;89:359–376.
153. Blasco MA. Telomeres and human disease: ageing, cancer and beyond. *Nat Rev Genet*. 2005;6:611–622.
154. Shay JW, Bacchetti S. A survey of telomerase activity in human cancer. *Eur J Cancer*. 1997;33:787–791.
155. Ulaner GA, Hu JF, Vu TH, Giudice LC, Hoffman AR. Telomerase activity in human development is regulated by human telomerase reverse transcriptase (hTERT) transcription and by alternate splicing of hTERT transcripts. *Cancer Res*. 1998;58:4168–4172.
156. Liu Y, Snow BE, Hande MP, et al. The telomerase reverse transcriptase is limiting and necessary for telomerase function in vivo. *Curr Biol*. 2000;10:1459–1462.
157. Blasco MA, Rizen M, Greider CW, Hanahan D. Differential regulation of telomerase activity and telomerase RNA during multi-stage tumorigenesis. *Nat Genet*. 1996;12:200–204.
158. Lu W, Zhang Y, Liu D, Songyang Z, Wan M. Telomeres-structure, function, and regulation. *Exp Cell Res*. 2013;319:133–141.
159. Tomas-Loba A, Flores I, Fernandez-Marcos PJ, et al. Telomerase reverse transcriptase delays aging in cancer-resistant mice. *Cell*. 2008;135:609–622.
160. Teng SC, Zakian VA. Telomere-telomere recombination is an efficient bypass pathway for telomere maintenance in Saccharomyces cerevisiae. *Mol Cell Biol*. 1999;19:8083–8093.
161. Henson JD, Neumann AA, Yeager TR, Reddel RR. Alternative lengthening of telomeres in mammalian cells. *Oncogene*. 2002;21:598–610.
162. Pardue ML, DeBaryshe PG. Retrotransposons provide an evolutionarily robust non-telomerase mechanism to maintain telomeres. *Annu Rev Genet*. 2003;37:485–511.
163. Reddel RR. Alternative lengthening of telomeres, telomerase, and cancer. *Cancer Lett*. 2003;194:155–162.
164. Counter CM, Avilion AA, LeFeuvre CE, et al. Telomere shortening associated with chromosome instability is arrested in immortal cells which express telomerase activity. *EMBO J*. 1992;11:1921–1929.
165. Bryan TM, Englezou A, Gupta J, Bacchetti S, Reddel RR. Telomere elongation in immortal human cells without detectable telomerase activity. *EMBO J*. 1995;14:4240–4248.
166. Dunham MA, Neumann AA, Fasching CL, Reddel RR. Telomere maintenance by recombination in human cells. *Nat Genet*. 2000;26:447–450.
167. Celli GB, Denchi EL, de Lange T. Ku70 stimulates fusion of dysfunctional telomeres yet protects chromosome ends from homologous recombination. *Nat Cell Biol*. 2006;8:885–890.
168. Palm W, Hockemeyer D, Kibe T, de Lange T. Functional dissection of human and mouse POT1 proteins. *Mol Cell Biol*. 2009;29:471–482.
169. Yeager TR, Neumann AA, Englezou A, Huschtscha LI, Noble JR, Reddel RR. Telomerase-negative immortalized human cells contain a novel type of promyelocytic leukemia (PML) body. *Cancer Res*. 1999;59:4175–4179.

170. Chung I, Osterwald S, Deeg KI, Rippe K. PML body meets telomere: the beginning of an ALTernate ending? *Nucleus*. 2012;3:263–275.
171. Roth CW, Kobeski F, Walter MF, Biessmann H. Chromosome end elongation by recombination in the mosquito Anopheles gambiae. *Mol Cell Biol*. 1997;17:5176–5183.
172. Mason JM, Frydrychova RC, Biessmann H. Drosophila telomeres: an exception providing new insights. *Bioessays*. 2008;30:25–37.
173. Dreesen O, Cross GA. Consequences of telomere shortening at an active VSG expression site in telomerase-deficient Trypanosoma brucei. *Eukaryot Cell*. 2006;5:2114–2119.
174. Dreesen O, Cross GA. Telomerase-independent stabilization of short telomeres in Trypanosoma brucei. *Mol Cell Biol*. 2006;26:4911–4919.
175. Dreesen O, Li B, Cross GA. Telomere structure and function in trypanosomes: a proposal. *Nat Rev Microbiol*. 2007;5:70–75.
176. Hovel-Miner GA, Boothroyd CE, Mugnier M, Dreesen O, Cross GA, Papavasiliou FN. Telomere length affects the frequency and mechanism of antigenic variation in Trypanosoma brucei. *PLoS Pathog*. 2012;8:e1002900.
177. Palm W, de Lange T. How shelterin protects mammalian telomeres. *Annu Rev Genet*. 2008;42:301–334.
178. Price CM, Boltz KA, Chaiken MF, Stewart JA, Beilstein MA, Shippen DE. Evolution of CST function in telomere maintenance. *Cell Cycle*. 2010;9:3157–3165.
179. Giraud-Panis MJ, Teixeira MT, Geli V, Gilson E. CST meets shelterin to keep telomeres in check. *Mol Cell*. 2010;39:665–676.
180. Marcand S, Gilson E, Shore D. A protein-counting mechanism for telomere length regulation in yeast. *Science*. 1997;275:986–990.
181. Ray A, Runge KW. The yeast telomere length counting machinery is sensitive to sequences at the telomere-nontelomere junction. *Mol Cell Biol*. 1999;19:31–45.
182. Smogorzewska A, van Steensel B, Bianchi A, et al. Control of human telomere length by TRF1 and TRF2. *Mol Cell Biol*. 2000;20:1659–1668.
183. Kanoh J, Ishikawa F. spRap1 and spRif1, recruited to telomeres by Taz1, are essential for telomere function in fission yeast. *Curr Biol*. 2001;11:1624–1630.
184. Neidle S, Parkinson GN. The structure of telomeric DNA. *Curr Opin Struct Biol*. 2003;13:275–283.
185. Lewis KA, Wuttke DS. Telomerase and telomere-associated proteins: structural insights into mechanism and evolution. *Structure*. 2012;20:28–39.
186. Schoeftner S, Blasco MA. Developmentally regulated transcription of mammalian telomeres by DNA-dependent RNA polymerase II. *Nat Cell Biol*. 2008;10:228–236.
187. Gray JT, Celander DW, Price CM, Cech TR. Cloning and expression of genes for the Oxytricha telomere-binding protein: specific subunit interactions in the telomeric complex. *Cell*. 1991;67:807–814.
188. Gottschling DE, Zakian VA. Telomere proteins: specific recognition and protection of the natural termini of Oxytricha macronuclear DNA. *Cell*. 1986;47:195–205.
189. Froelich-Ammon SJ, Dickinson BA, Bevilacqua JM, Schultz SC, Cech TR. Modulation of telomerase activity by telomere DNA-binding proteins in Oxytricha. *Genes Dev*. 1998;12:1504–1514.
190. Theobald DL, Cervantes RB, Lundblad V, Wuttke DS. Homology among telomeric end-protection proteins. *Structure*. 2003;11:1049–1050.
191. Xin H, Liu D, Wan M, et al. TPP1 is a homologue of ciliate TEBP-beta and interacts with POT1 to recruit telomerase. *Nature*. 2007;445:559–562.
192. Gao H, Cervantes RB, Mandell EK, Otero JH, Lundblad V. RPA-like proteins mediate yeast telomere function. *Nat Struct Mol Biol*. 2007;14:208–214.
193. Evans SK, Lundblad V. Positive and negative regulation of telomerase access to the telomere. *J Cell Sci*. 2000;113(Pt 19):3357–3364.

194. Faure V, Coulon S, Hardy J, Geli V. Cdc13 and telomerase bind through different mechanisms at the lagging- and leading-strand telomeres. *Mol Cell.* 2010;38:842–852.
195. Schramke V, Luciano P, Brevet V, et al. RPA regulates telomerase action by providing Est1p access to chromosome ends. *Nat Genet.* 2004;36:46–54.
196. Chen LY, Redon S, Lingner J. The human CST complex is a terminator of telomerase activity. *Nature.* 2012;488:540–544.
197. Miyake Y, Nakamura M, Nabetani A, et al. RPA-like mammalian Ctc1-Stn1-Ten1 complex binds to single-stranded DNA and protects telomeres independently of the Pot1 pathway. *Mol Cell.* 2009;36:193–206.
198. Horvath MP, Schweiker VL, Bevilacqua JM, Ruggles JA, Schultz SC. Crystal structure of the Oxytricha nova telomere end binding protein complexed with single strand DNA. *Cell.* 1998;95:963–974.
199. Wold MS. Replication protein A: a heterotrimeric, single-stranded DNA-binding protein required for eukaryotic DNA metabolism. *Annu Rev Biochem.* 1997;66:61–92.
200. Siqueira-Neto JL, Lira CB, Giardini MA, et al. Leishmania replication protein A-1 binds in vivo single-stranded telomeric DNA. *Biochem Biophys Res Commun.* 2007;358:417–423.
201. Sakaguchi K, Ishibashi T, Uchiyama Y, Iwabata K. The multi-replication protein A (RPA) system—a new perspective. *FEBS J.* 2009;276:943–963.
202. Da Silveira Rde C, Da Silva MS, Nunes VS, Perez AM, Cano MI. The natural absence of RPA1N domain did not impair Leishmania amazonensis RPA-1 participation in DNA damage response and telomere protection. *Parasitology.* 2013;140:547–559.
203. Shore D, Bianchi A. Telomere length regulation: coupling DNA end processing to feedback regulation of telomerase. *EMBO J.* 2009;28:2309–2322.
204. Konig P, Giraldo R, Chapman L, Rhodes D. The crystal structure of the DNA-binding domain of yeast RAP1 in complex with telomeric DNA. *Cell.* 1996;85:125–136.
205. Marcand S, Brevet V, Gilson E. Progressive cis-inhibition of telomerase upon telomere elongation. *EMBO J.* 1999;18:3509–3519.
206. Krauskopf A, Blackburn EH. Control of telomere growth by interactions of RAP1 with the most distal telomeric repeats. *Nature.* 1996;383:354–357.
207. Cooper JP, Nimmo ER, Allshire RC, Cech TR. Regulation of telomere length and function by a Myb-domain protein in fission yeast. *Nature.* 1997;385:744–747.
208. Broccoli D, Smogorzewska A, Chong L, de Lange T. Human telomeres contain two distinct Myb-related proteins, TRF1 and TRF2. *Nat Genet.* 1997;17:231–235.
209. Ye JZ, Donigian JR, van Overbeek M, et al. TIN2 binds TRF1 and TRF2 simultaneously and stabilizes the TRF2 complex on telomeres. *J Biol Chem.* 2004;279:47264–47271.
210. Liu D, O'Connor MS, Qin J, Songyang Z. Telosome, a mammalian telomere-associated complex formed by multiple telomeric proteins. *J Biol Chem.* 2004;279:51338–51342.
211. Lei M, Podell ER, Cech TR. Structure of human POT1 bound to telomeric single-stranded DNA provides a model for chromosome end-protection. *Nat Struct Mol Biol.* 2004;11:1223–1229.
212. Loayza D, Parsons H, Donigian J, Hoke K, de Lange T. DNA binding features of human POT1: a nonamer 5'-TAGGGTTAG-3' minimal binding site, sequence specificity, and internal binding to multimeric sites. *J Biol Chem.* 2004;279:13241–13248.
213. Stern JL, Bryan TM. Telomerase recruitment to telomeres. *Cytogenet Genome Res.* 2008;122:243–254.
214. Wu L, Multani AS, He H, et al. Pot1 deficiency initiates DNA damage checkpoint activation and aberrant homologous recombination at telomeres. *Cell.* 2006;126:49–62.

215. Hsu HL, Gilley D, Galande SA, et al. Ku acts in a unique way at the mammalian telomere to prevent end joining. *Genes Dev.* 2000;14:2807–2812.
216. Ribes-Zamora A, Mihalek I, Lichtarge O, Bertuch AA. Distinct faces of the Ku heterodimer mediate DNA repair and telomeric functions. *Nat Struct Mol Biol.* 2007;14:301–307.
217. Lopez CR, Ribes-Zamora A, Indiviglio SM, Williams CL, Haricharan S, Bertuch AA. Ku must load directly onto the chromosome end in order to mediate its telomeric functions. *PLoS Genet.* 2011;7:e1002233.
218. Arora R, Brun CM, Azzalin CM. TERRA: long noncoding RNA at eukaryotic telomeres. *Prog Mol Subcell Biol.* 2011;51:65–94.
219. Iglesias N, Redon S, Pfeiffer V, Dees M, Lingner J, Luke B. Subtelomeric repetitive elements determine TERRA regulation by Rap1/Rif and Rap1/Sir complexes in yeast. *EMBO Rep.* 2011;12:587–593.
220. Farnung BO, Giulotto E, Azzalin CM. Promoting transcription of chromosome ends. *Transcription.* 2010;1:140–143.
221. Farnung BO, Brun CM, Arora R, Lorenzi LE, Azzalin CM. Telomerase efficiently elongates highly transcribing telomeres in human cancer cells. *PLoS One.* 2012;7: e35714.
222. Pfeiffer V, Lingner J. TERRA promotes telomere shortening through exonuclease 1-mediated resection of chromosome ends. *PLoS Genet.* 2012;8:e1002747.
223. Redon S, Zemp I, Lingner J. A three-state model for the regulation of telomerase by TERRA and hnRNPA1. *Nucleic Acids Res.* 2013;41:9117–9128.
224. Ishikawa F, Matunis MJ, Dreyfuss G, Cech TR. Nuclear proteins that bind the pre-mRNA 3' splice site sequence r(UUAG/G) and the human telomeric DNA sequence d(TTAGGG)n. *Mol Cell Biol.* 1993;13:4301–4310.
225. Flynn RL, Centore RC, O'Sullivan RJ, et al. TERRA and hnRNPA1 orchestrate an RPA-to-POT1 switch on telomeric single-stranded DNA. *Nature.* 2011;471:532–536.
226. Arnoult N, Van Beneden A, Decottignies A. Telomere length regulates TERRA levels through increased trimethylation of telomeric H3K9 and HP1alpha. *Nat Struct Mol Biol.* 2012;19:948–956.
227. Hayflick L. The limited in vitro lifetime of human diploid cell strains. *Exp Cell Res.* 1965;37:614–636.
228. McEachern MJ, Haber JE. Break-induced replication and recombinational telomere elongation in yeast. *Annu Rev Biochem.* 2006;75:111–135.
229. Kong CM, Lee XW, Wang X. Telomere shortening in human diseases. *FEBS J.* 2013;280:3180–3193.
230. Mason PJ, Wilson DB, Bessler M. Dyskeratosis congenita—a disease of dysfunctional telomere maintenance. *Curr Mol Med.* 2005;5:159–170.
231. Guo X, Deng Y, Lin Y, et al. Dysfunctional telomeres activate an ATM-ATR-dependent DNA damage response to suppress tumorigenesis. *EMBO J.* 2007;26:4709–4719.
232. Denchi EL, de Lange T. Protection of telomeres through independent control of ATM and ATR by TRF2 and POT1. *Nature.* 2007;448:1068–1071.
233. Flynn RL, Chang S, Zou L. RPA and POT1: friends or foes at telomeres? *Cell Cycle.* 2012;11:652–657.
234. Takai H, Smogorzewska A, de Lange T. DNA damage foci at dysfunctional telomeres. *Curr Biol.* 2003;13:1549–1556.
235. Dimitrova N, Chen YC, Spector DL, de Lange T. 53BP1 promotes non-homologous end joining of telomeres by increasing chromatin mobility. *Nature.* 2008;456:524–528.
236. Counter CM, Meyerson M, Eaton EN, et al. Telomerase activity is restored in human cells by ectopic expression of hTERT (hEST2), the catalytic subunit of telomerase. *Oncogene.* 1998;16:1217–1222.

237. Cerone MA, Londono-Vallejo JA, Bacchetti S. Telomere maintenance by telomerase and by recombination can coexist in human cells. *Hum Mol Genet.* 2001;10:1945–1952.
238. Mangahas JL, Alexander MK, Sandell LL, Zakian VA. Repair of chromosome ends after telomere loss in Saccharomyces. *Mol Biol Cell.* 2001;12:4078–4089.
239. Pennaneach V, Putnam CD, Kolodner RD. Chromosome healing by de novo telomere addition in Saccharomyces cerevisiae. *Mol Microbiol.* 2006;59:1357–1368.
240. Smogorzewska A, Karlseder J, Holtgreve-Grez H, Jauch A, de Lange T. DNA ligase IV-dependent NHEJ of deprotected mammalian telomeres in G1 and G2. *Curr Biol.* 2002;12:1635–1644.
241. Mieczkowski PA, Mieczkowska JO, Dominska M, Petes TD. Genetic regulation of telomere-telomere fusions in the yeast Saccharomyces cerevisiae. *Proc Natl Acad Sci USA.* 2003;100:10854–10859.
242. McEachern MJ. Telomeres: guardians of genomic integrity or double agents of evolution? In: Nosek J, Tomáska L, eds. *Origin and Evolution of Telomeres.* Landes: Bioscience; 2008:100–108.

CHAPTER TWO

Human Telomeres and Telomere Biology Disorders

Sharon A. Savage
Division of Cancer Epidemiology and Genetics, National Cancer Institute, National Institutes of Health, Bethesda, Maryland, USA

Contents

Abstract

Telomeres consist of long nucleotide repeats and a protein complex at chromosome ends essential for chromosome stability. Telomeres shorten with each cell division and thus are markers of cellular age. Dyskeratosis congenita (DC) is a cancer-prone inherited bone marrow failure syndrome caused by germ-line mutations in key

Progress in Molecular Biology and Translational Science
Volume 125, ISSN 1877-1173
http://dx.doi.org/10.1016/B978-0-12-397898-1.00002-5

telomere biology genes that result in extremely short telomeres. The triad of nail dysplasia, abnormal skin pigmentation, and oral leukoplakia is diagnostic of DC but highly variable. Patients with DC may also have but numerous other medical problems, including pulmonary fibrosis, liver abnormalities, avascular necrosis of the hips, and stenosis of the esophagus, lacrimal ducts, and/or urethra. All modes of inheritance have been reported in DC and *de novo* mutations are common. Broad phenotypic heterogeneity occurs within DC. Clinically severe variants of DC are Hoyeraal–Hreidarsson syndrome and Revesz syndrome. Coats plus syndrome joined the spectrum of DC with the discovery that it is caused by mutations in a telomere-capping gene. Less clinically severe variants, such as subsets of apparently isolated aplastic anemia or pulmonary fibrosis, have also been recognized. These patients may not have the DC-associated mucocutaneous triad or complicated medical features, but they do have the same underlying genetic etiology. This has led to the use of the descriptive term telomere biology disorder (TBD). This chapter will review the connection between telomere biology and human disease through the examples of DC and its related TBDs.

1. INTRODUCTION

Over the last two decades, aberrations in telomere biology have emerged as an important cause of disease in humans. Telomeres consist of long hexameric nucleotide repeats and a protein complex at chromosome ends. They are critical for the maintenance of chromosomal integrity. Dyskeratosis congenita (DC), the prototypical telomere biology disorder (TBD), is also a cancer-prone inherited bone marrow failure (BMF) syndrome. Patients with DC are classically diagnosed by the presence of a mucocutaneous triad that includes nail dysplasia, abnormal skin pigmentation, and oral leukoplakia. Numerous other medical problems occur in DC, such as BMF, pulmonary fibrosis, liver abnormalities, and elevated risk of certain cancers. Patients with DC have extremely short telomeres for their age and germ-line mutations in key telomere biology genes. It is now appreciated that a clinical spectrum of disorders are caused by mutations in the same genes that cause DC, for example, individuals with apparently isolated pulmonary fibrosis or BMF. These patients may not have the DC-associated mucocutaneous triad or complicated medical features, but do have the same underlying genetic etiology, hence the creation of the descriptive term to unite seemingly different illnesses: TBDs. This chapter will review the connection between telomere biology and human disease through the example of DC and its related TBDs.

2. TELOMERE BIOLOGY IN HUMANS

2.1. Telomeric structure

Telomeres are nucleoprotein complexes located at the ends of all eukaryotic chromosomes that consist of long, tandem TTAGGG nucleotide repeats and a protein complex.[1–3] The nucleotide repeats fold back to create a T-loop structure to which numerous proteins bind to and interact with in order to protect the ends of chromosomes and maintain genome integrity.[1–3] The duplex portion of human telomeric repeat sequences ranges from two to over 14 kb in length. At the end of this sequence, there is a single-strand DNA overhang of approximately 200 nucleotides.[4]

The telomeric sequence at chromosome ends shortens with each cell division due to the inability of DNA polymerase to fully replicate the 3′ end of DNA sequences.[5,6] This "end-replication" problem eventually results in telomeres reaching a critically short length due to the loss of telomere end protection.[7] Consequently, cell-cycle arrest is triggered that leads to cellular senescence or apoptosis.[8] Cells can bypass these processes through the upregulation of the telomerase p53 or the alternative lengthening of telomeres (ALT) pathway, which allows for continued cell division in the setting of genomic instability and the development of somatic mutations that can lead to carcinogenesis.[9,10]

2.2. Telomerase and telomere-associated proteins

The cellular end-replication problem is primarily addressed by telomerase, a specialized ribonucleoprotein, which adds repetitive G-rich sequence on to telomere ends.[11] The telomerase enzyme and its components are highly evolutionarily conserved. It consists of two core components: TERT, a reverse transcriptase, and TERC, an RNA that contains the template for telomere repeat addition.[12] TERT and TERC are sufficient for telomerase reconstitution *in vitro*, but *in vivo* telomerase biogenesis, localization, and activity require additional factors. Dyskerin (*DKC1*) associates with TERC and other H/ACA box-containing small nucleolar RNAs (snoRNAs) and is required for normal TERC levels and telomerase activity *in vivo*.[13] Dyskerin (encoded by *DKC1*) forms a ribonucleoprotein complex with NOP10, NHP2, and GAR1, which are also found in a complex with TERC. This complex is important in the stability and regulation of telomerase.[14,15] Additionally, dyskerin is involved in posttranscriptional pseudouridylation; this connection to the shortened telomeres seen as a consequence of mutant *DKC1* is not thoroughly understood.[16–18] The assembly of the catalytically active telomerase complex also requires NAF1, which is exchanged for GAR1 in the dyskerin/NOP10/NHP2 complex.[19] Lastly, there is the protein TCAB1 (encoded by *WRAP53*), which associates with active telomerase by binding to the CAB box motif of TERC and is responsible for the localization of telomerase to nuclear Cajal bodies, a critical step for telomere maintenance.[20]

Telomerase activity is carefully regulated in all tissues. It is typically expressed only during early embryogenesis.[21] From the neonatal period onward, telomerase activity is largely repressed, except in certain highly proliferative organs such as skin, intestine, and bone marrow, which are thought to contain stem cell-like subpopulations, and in dividing lymphocytes, ovaries, and testes.[21–24] Telomerase is also upregulated in most cancer cells, reflecting the need for telomere maintenance for proliferative potential.[10,25]

2.3. Alternative lengthening of telomeres

Carcinogenesis requires mechanisms that allow cells to divide despite the presence of exceedingly short telomeres. Upregulation of telomerase occurs in 80–90% of somatic cancer tissues.[10,25] This can lead to significant telomere length (TL) heterogeneity within cancers but suffices to help maintain cellular proliferation. The ALT pathway is a homologous recombination (HR)-mediated mechanism, which involves copying of the telomeric DNA template.[26,27] This causes TL heterogeneity, extrachromosomal linear and circular telomeric DNA, increased telomere-sister chromatid exchange, and the presence of promyelocytic leukemia bodies, containing telomeric DNA, telomere-associated proteins, and HR factors. Although ALT is present in some somatic cancer cells, normal human tissues have not yet been reported to use ALT, but a recent study suggests that normal murine somatic tissues may use ALT.[9]

2.4. Measuring TL

There are several methods used in the research setting to determine TL in whole cells and in total DNA preparations.[28,29] Terminal restriction fragment (TRF) measurement on Southern blots is the most commonly used method of TL determination. It uses restriction enzymes to digest telomeric DNA, and the resultant DNA fragments are run on a Southern blot to quantify the TL. The TRF method is useful in the research setting. However, it requires several hundred nanograms of high-quality DNA and requires quantification of a smear of DNA fragments.

Fluorescence *in situ* hybridization (FISH) with telomere-specific probes is commonly used to measure TL on metaphase chromosomes and on fixed tissues. The fluorescent intensity of the telomeric probe is compared with that of another DNA-specific probe. This approach is useful in comparing TL via signal intensity in individual chromosomes or cells but is time-intensive.[28,29] Single telomere length analysis is a PCR-based technique that determines TL of single chromosomes. It uses PCR primers that include but the telomeric and subtelomeric DNA sequence and thus is specific for a limited number of chromosome arms.[30,31]

Since its development, the quantitative PCR (qPCR) method of measuring relative TL has been widely used in epidemiology studies.[32,33] This high-throughput platform requires small amounts of high-quality DNA and determines a ratio of telomere signal to single-copy gene signal (T/S ratio). The correlation between TL measured by Southern blot and relative TL by qPCR varies by study with a R^2 range of 0.27–0.84.[32–35] The use of qPCR TL measurement and epidemiology studies is further reviewed in Chapter 5 of this book.

The only clinically validated method of TL measurement, to date, determines TL in white blood cell (WBC) subsets through automated multicolor flow cytometry with FISH (flow FISH).[36] This method requires fresh or cryopreserved blood samples but gives specific TL measurements on WBC subsets. Flow FISH is highly sensitive and specific for differentiating patients with DC and related TBDs from their unaffected relatives, from patients with other inherited bone marrow failure syndromes (IBMFS) and from healthy controls.[37,38] It is important to note that measurement of TL by other methods, including qPCR and TRF, is used in research studies of DC but the diagnostic sensitivity and specificity of these methods are not known.[39,40]

3. WHAT IS A TBD?

Abnormalities in telomere biology causing clinically significant diseases were first recognized in patients with DC.[41,42] The connection between diseases with isolated clinical features seen in DC and telomere biology was made first through studies of individuals with severe aplastic anemia[43,44] and then in those with pulmonary fibrosis.[45,46] Additional reports of germ-line mutations in telomere biology genes associated with disease in individuals with liver cirrhosis of unknown etiology[47] and in those with Coats plus syndrome[48,49] further expanded the clinical phenotypes of DC. This led to the use of the descriptive umbrella term: TBDs.

4. CLINICAL FEATURES OF THE TBDs

The TBDs are a set of complex illnesses related to aberrant telomere biology. They range from clinically very severe diseases with multisystem involvement, as in DC, to diseases with only one organ system affected, such as pulmonary fibrosis. Clinically silent carriers of a TBD-associated genetic mutation have also been reported. TBDs should be considered as a spectrum of related disorders united by a common biology. For example, a child with the classic mucocutaneous triad of DC and BMF may not seem clinically similar to a middle-aged adult with pulmonary fibrosis, but they may have a germ-line mutation in the same gene, and both may have an elevated risk of DC-associated cancer.

4.1. Dyskeratosis congenita

DC-associated mucocutaneous features were first described in the early 1900s.[50,51] Initially, it was designated Zinsser–Cole–Engman syndrome based on the authors of the clinical reports. The name appears to have later been changed to DC because of the mucocutaneous features and its congenital nature. During the mid- to late twentieth century, additional clinical reports expanded the phenotype as a greater appreciation of its multisystem and non-gender-specific nature.

The classic triad of nail dystrophy, lacy reticular pigmentation of neck/upper chest, and oral leukoplakia is diagnostic (Fig. 2.1). However, due to variable expressivity and/or incomplete penetrance, the classic triad is not

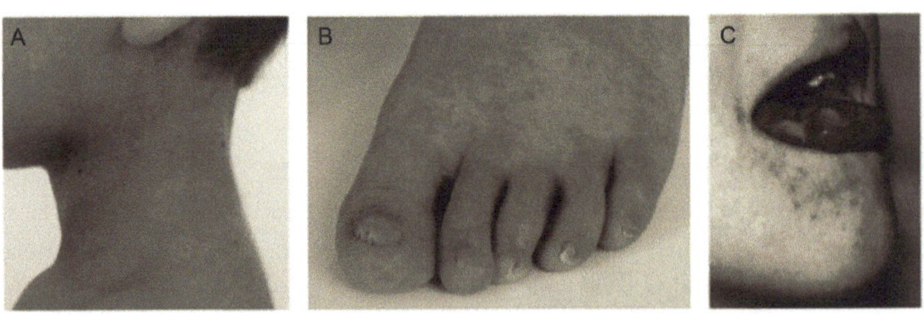

Figure 2.1 *The diagnostic triad of dyskeratosis congenita.* (A) Skin hyper- and hypopigmentation; (B) toenail dystrophy; (C) oral leukoplakia. (See the color plate.)

always present in all individuals with DC or may worsen with age and may thus be overlooked when a patient presents with other medical problems, such as severe BMF. Patients with DC may have other significant medical problems.[52–57] Stenosis of the esophagus with or without webbing may occur and require dilation. Urethral stenosis in males can also occur and is also treated with dilation. Pulmonary fibrosis and nonalcoholic/noninfectious liver disease may be a significant challenge in DC and are discussed further below. Additionally, bone abnormalities have been reported. Avascular necrosis of the hips or shoulders may occur and cause significant pain. Several patients in the National Cancer Institute's (NCI's) DC cohort (ClinicalTrials.gov Identifier NCT00027274) have required hip replacements in their 20s due to avascular necrosis (S. Savage, unpublished data). Patients with DC may also have osteopenia and increased risk of bone fractures. Recognizing the potential for these problems can help guide orthopedic therapy.

The dental problems in DC may include a decreased root/crown ratio, taurodontism, increased dental caries, hypodontia, thin enamel structure, periodontitis, tooth loss, and blunted roots.[58] A study of the ocular manifestations of DC in the NCI cohort reported that 28% of DC patients have an obstructed lacrimal drainage system.[59] Patients were also reported to have entropion and trichiasis, possibly secondary to epithelial abnormalities in the ocular skin and mucous membranes. Revesz syndrome (RS), a severe subtype of DC discussed below, is also marked by bilateral exudative retinopathy.

Recent evidence from the NCI DC cohort suggests that patients with DC may have higher rates of neuropsychiatric disorders than the general population.[60] Fifty percent of children and 75% of adults with a DC or DC-like diagnosis had experienced a psychiatric disorder, compared with 25% of chronically ill children.[60] DC patients had evidence of psychiatric disorders (mood, anxiety, psychotic, and adjustment disorders) or neurocognitive disorders (attention deficit/hyperactivity disorder, intellectual disability, learning disabilities, and pervasive developmental disorders). However, larger sample sizes are needed to better understand the findings of this study.

Patients with DC are at increased risk of cancer. Early case reports included the co-occurrence of MDS, AML, or head and neck squamous cell cancer in patients with DC. Other cancers, including lymphoma and cancers of the gastrointestinal tract and liver, are rare but have been reported in DC. To date, the only quantitative study of cancer in DC consists of a systematic literature review combined with the data from the NCI's DC cohort study. It found an overall 11-fold increased risk of cancer.[61] This study found that patients with

DC are at a 195-fold increased risk of AML, a 2663-fold greater risk of myelodysplastic syndrome, and an eightfold greater risk of solid tumors (including an 1154-fold increase for tongue cancer) compared with the general population. The risk was increased 43-fold for cervical squamous cell carcinoma and 34-fold for non-Hodgkin's lymphoma.[61] In sum, the actuarial risk of any cancer in DC is 40% by age 50. It is important to note that although this study provided the first statistical data on the rates of cancer in DC, it is limited by sample size and by its retrospective nature. Reporting bias of cases of DC with cancer in the literature is likely, and this would skew the data toward higher rates. However, even with those considerations, it is clear that patients with DC are at a significant risk of cancers that tend to affect tissues with high turnover. Early detection through regular clinical examinations is important to reduce cancer-associated mortality in this population.

4.2. Hoyeraal–Hreidarsson syndrome

In 1970, Hoyeraal *et al.* reported the occurrence of intrauterine growth retardation (IUGR), microcephaly, cerebellar hypoplasia, and thrombocytopenia in two brothers.[62] An unrelated affected male with similar features was reported in 1988 by Hreidarsson *et al.*[63] Additional cases were reported over the next two decades that also noted the presence of immunodeficiency in this syndrome.[57,64] Nonspecific enteropathies have also been reported in Hoyeraal–Hreidarsson (HH). The DC mucocutaneous triad develops in many HH patients, but may not be present in infancy. The presence of cerebellar hypoplasia in the setting of DC-associated features is generally required for the diagnosis of HH (Fig. 2.2A). The connection between DC and HH was suggested due to overlapping clinical features and was confirmed when

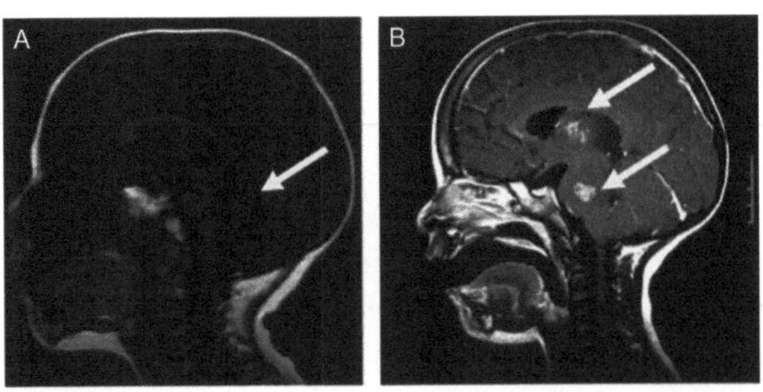

Figure 2.2 *Neurological findings in telomere biology disorders.* (A) Cerebellar hypoplasia in a patient with Hoyeraal–Hreidarsson syndrome; (B) intracranial calcifications in a patient with Coats plus syndrome. *Photo 2.2(B) Courtesy of Dr. Yanick Crow.*

mutations in *DKC1* were identified as causes in both disorders.[41,65] Mutations in *TINF2* and AR mutations in *TERT*, *NHP2*, *NOP10*, and *RTEL1* have also been reported as causes of HH.[66–71]

4.3. Revesz syndrome

Revesz *et al.* reported a case of a 6-month-old infant with bilateral exudative retinopathy (bilateral Coats disease) who went on to develop severe BMF in 1992.[72] This case and subsequent reports of similar cases noted that patients with RS had IUGR; intracranial calcifications; developmental delay; fine, sparse hair; and nail dystrophy. The clinical overlap with DC suggested a common etiology. This hypothesis was confirmed with the discovery that patients with RS have very short telomeres and many have germ-line mutations in *TINF2*, a DC-associated gene.[68] The specific diagnosis of RS requires the identification of bilateral exudative retinopathy, which must be distinguished from proliferative retinopathy, reported in non-RS DC.[59]

4.4. Coats plus syndrome/CRMCC

Coats plus syndrome, also known as cerebroretinal microangiopathy with calcification and cysts (CRMCC), joined the cadre of DC-related TBDs with the discovery of autosomal recessive (AR) compound heterozygous mutations in *CTC1*, a key telomere-capping gene.[48,49,73] Coats plus syndrome patients have bilateral exudative retinopathy, retinal telangiectasias, IUGR, intracranial calcifications, bone abnormalities with poor healing, and gastrointestinal vascular ectasias (Fig. 2.3). Some Coats plus syndrome patients have been reported to have features seen in DC including dystrophic nails, sparse or graying hair, and anemia. Additionally, the intracranial calcifications (Fig. 2.2B) and bilateral exudative retinopathy overlap with RS.[72] Mutations in *CTC1* appear to account for the majority of Coats plus syndrome cases. In addition, Coats plus syndrome patients have telomeres that are below the first percentile for age; heterozygous mutation carriers have TLs that are below average.[48]

4.5. Aplastic anemia

The etiology of aplastic anemia or BMF is multifactorial with both inherited and acquired forms. Acquired aplastic anemia is often immune-mediated and can be related to environmental exposures, infections, or idiosyncratic reactions to medications. Patients with an inherited BMF may develop aplastic anemia as their first presenting sign. This has been reported to occur

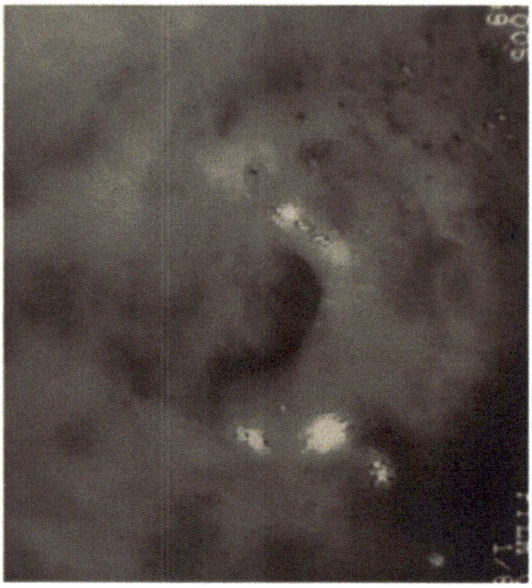

Figure 2.3 Gastrointestinal vascular ectasias in Coats plus syndrome. *Courtesy of Dr. Yanick Crow.*

in patients with DC, as well as Fanconi anemia, Shwachman–Diamond syndrome, and other inherited BMFs. Differentiating acquired BMFs from inherited BMFs is critical for patient management since acquired aplastic anemia often responds to immunosuppressive therapy, whereas the inherited forms do not. Additionally, hematopoietic stem cell transplantation (HSCT) regimens need to be specifically tailored for the disorder.

Approximately 10% of patients with apparent isolated aplastic anemia have AD mutations in *TERC* or *TERT*.[43,44] In these cases, TL is usually less than the 10th percentile for age.[74] Detailed review of the family history of these patients may review disorders also seen in DC, such as pulmonary fibrosis, mild cytopenias, leukemia, and squamous cell cancer. These patients should be considered to have a milder variant of DC and thus are often classified as having a TBD. The long-term risk of DC-associated medical complications in patients with isolated BMF and *TERT* or *TERC* mutations is not known, but the later development of the mucocutaneous triad, pulmonary fibrosis, and liver disease in these patients is possible.

4.6. Pulmonary fibrosis

Approximately 5–10% of patients with apparently isolated pulmonary fibrosis have been found to have a TBD due to AD germ-line *TERC* or *TERT* mutations.[45,46,75,76] Idiopathic pulmonary fibrosis is a complex, multifactorial disease that leads to progressive lung scarring and fibrotic changes. The majority of cases are sporadic with a limited number of known environmental risk factors. However, a detailed family history often reveals features of a DC-related TBD, such as aplastic anemia, or nonalcoholic/noninfectious liver disease. Similar to idiopathic pulmonary fibrosis, the optimal clinical management of pulmonary fibrosis due to aberrant telomere biology is not yet known.

4.7. Nonalcoholic/noninfectious liver disease

Hepatic fibrosis, noncirrhotic portal hypertension, and hepatopulmonary syndrome have been reported as DC-associated complications.[52,53] These complications led to a study by Calado *et al.* of five families with liver disease in combination with hematologic and autoimmune disorders.[47] This study connected loss of function *TERC* or *TERT* mutations with the clinical phenotypes. It is important to note that in this study, nonalcoholic/noninfectious liver disease was present as an isolated finding in individuals with mutations and relatives with BMF who had the same mutation. A subsequent study of individuals with cirrhosis of unknown etiology found that these patients more commonly had rare mutations in *TERT* or *TERC* compared with controls (3.7% vs. 0.8%).[77] This suggests that a subset of patients with liver disease of unknown etiology may have an underlying TBD and need to be treated as such.

4.8. Genetic anticipation

As noted throughout this chapter, the penetrance, severity, and time of onset of the clinical features of TBDs are variable, even among family members with the same germ-line mutations. Disease anticipation, the occurrence of increasing disease severity and earlier onset with successive generations, has been observed in several multigenerational families with AD DC.[55,78–80] The earlier age at onset has been associated with progressive telomere shortening with each successive generation. This concept is an especially important genetic counseling concept to include in discussions with families. It is possible that a clinically healthy mutation-carrier parent could have a child with clinical features of a TBD or even a severe form of DC.[67]

5. GERM-LINE GENETICS OF TBDs

DC is inherited in X-linked, AD or AR patterns.[53,81] *De novo* germ-line mutations are also relatively frequent in DC. Individuals with less severe clinical complications of a TBD may have a germ-line mutation in one of the same genes that causes DC that is due to variable penetrance of the phenotype and/or variable expressivity of the disease-associated mutation.

To date, about 70% of classic DC patients have an identifiable germ-line mutation.[53,81] These mutations occur in genes responsible for the functioning and maintenance of telomeres (Fig. 2.4). Currently, there are nine known DC-associated genes (*DKC1*, *TERT*, *TERC*, *TINF2*, *WRAP53*, *NOP10*, *NHP2*, *CTC1*, and *RTEL1*) (Table 2.1).[53,81]

X-linked inheritance of germ-line mutations in dyskerin (encoded by *DKC1*) was the first proven genetic cause of DC.[41,88,89] *DKC1*-mutant patient-derived fibroblasts were shown to have very short telomeres.[89] These cells have reduced telomerase activity because *TERC* levels are lower in DKC1-mutant cells. The connection between DC and telomere biology was further solidified when genetic linkage analysis of a large DC family found AD mutations in *TERC*.[82] Subsequently, a combination of linkage[79] and candidate gene sequencing found mutations in *TERT*, as a cause of DC and in cases of isolated aplastic anemia and pulmonary fibrosis.[44,83,90,91]

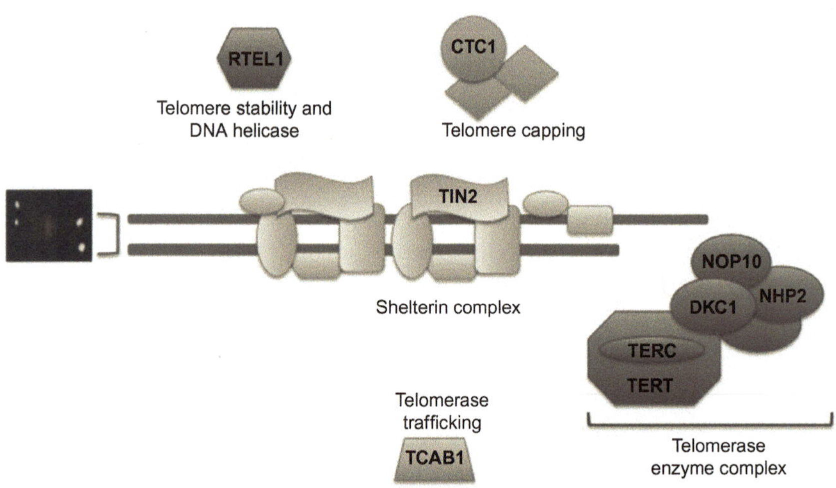

Figure 2.4 *Schematic of the telomere and functions of the proteins affected in dyskeratosis congenita and the related telomere biology disorders*. Protein names are shown. TCAB1, telomere Cajal body-associated protein 1 (gene name: *WRAP53*); TIN2, TRF1-interacting nuclear factor 2 (*TINF2*); NOP10, NOP10 ribonucleoprotein (*NOP10*); NHP2, NHP2 ribonucleoprotein (*NHP2*); DKC1, dyskerin (*DKC1*); TERC, telomerase RNA component (*TERC*); TERT, telomerase (*TERT*); RTEL1, regulator of telomere elongation helicase 1 (*RTEL1*); CTC1, CTS telomere maintenance complex component 1 (*CTC1*).

Table 2.1 Known genetic causes of telomere biology disorders

***Gene*, protein name(s)**	**Chromosomal locus**	**Year reported**	**MIM #**	**Inheritance**	**Disorder(s)**
DKC1, DKC1, dyskerin	Xq28	1998[42]	300126	XLR	DC, HH
TERC, TERC, telomerase RNA component	3q26.3	2001[82]	602322	AD	DC, SAA, PF, LD
TERT, TERT, telomerase reverse transcriptase	5p15.53	2005[44,83]	187270	AD, AR	DC, SAA, PF, FLD, AML, HH (AR)
NOP10, NOP10, NOLA3, nucleolar protein family A, member 3	15q14–q15	2007[70]	606471	AR	DC
TINF2, TIN2, TERF1 (TRF1)-interacting nuclear factor 2	14q11.2	2008[68]	604319	AD	DC, HH
NHP2, NHP2, NOLA2 nucleolar protein family A, member 2	5q35.5	2008[71]	606470	AR	DC
WRAP53, WD repeat-containing protein antisense to TP53; TCAB1, telomerase Cajal body protein 1	17p13.1	2011[84]	612661	AR	DC
CTC1, CTC1, conserved telomere maintenance component 1	17p13.1	2012[48,49,85]	613129	AR	DC, CP
RTEL1, RTEL1, regulator of telomere elongation helicase 1	20q13.33	2013[67,86,87]	608833	AD, AR	DC, HH

XLR, X-linked recessive; AD, autosomal dominant; AR autosomal recessive; DC, dyskeratosis congenita; RS, Revesz syndrome; HH, Hoyeraal–Hreidarsson syndrome; CP, Coats plus syndrome; SAA, severe aplastic anemia; PF, pulmonary fibrosis; LD, fibrotic liver disease; AML, acute myelogenous leukemia.
Online Mendelian Inheritance in Man (MIM) number derived from http://omim.org/.

These findings united the seemingly disparate diseases under a common etiology of DC-associated TBDs.[53,55]

Additional connections between DC and telomerase were made with the discovery of AR mutations in *NOP10* as a cause of DC. This finding was possible through the use of homozygosity mapping in a consanguineous family with DC.[70] Subsequent candidate gene sequencing of other components of the telomerase enzyme complex identified AR mutations of *NHP2* in two families.[71]

Compound heterozygous mutations in TCAB1 (encoded by *WRAP53)* were reported to cause DC in two families.[84] TCAB1 is an essential component of the telomerase holoenzyme complex because it is required from telomerase trafficking to Cajal bodies. DC-associated mutations in TCAB1 caused abnormal telomerase RNA localization, thus preventing telomere elongation by telomerase.

Key telomere biology proteins that are not part of the telomerase holoenzyme complex are also involved in the etiology of DC. This was first shown with the discovery of AD mutations in *TINF2* in a linkage mapping study of a large family with DC.[68] *TINF2* encodes the TIN2 protein, a key component of the six-protein shelterin telomere protection complex. Subsequent studies confirmed AD *TINF2* mutations as a relatively common cause of DC, accounting for about 11–20% of cases.[68,69,92]

Aberrations in telomere capping and protection were further shown to cause human disease with the discovery of compound heterozygous mutations in the telomere-capping protein encoded by *CTC1.*[48,49] Coats plus syndrome (CRMCC) and DC were united by biology when whole-exome sequencing identified that compound heterozygous mutations in the telomere-capping protein encoded by CTC1 can cause both disorders (Table 2.1 and Fig. 2.1). Patients with these mutations have short telomeres and features that phenotypically overlapped with DC.[73] Subsequently, candidate gene sequence analysis has led to the discovery of AR CTC1 mutations in patients with DC.[85,93]

The discovery of mutations in *RTEL1* further expanded the biology underlying DC and the TBDs. Several groups performed whole-exome sequencing in patients with DC and their families, which showed that germ-line *RTEL1* mutations could cause DC and related TBDs.[67,86,87] Most of the *RTEL1* mutations appear to be AR, but AD mutations have been reported (e.g., BMF in a proband with a sister who had DC-associated cancer).[67] The RTEL1 protein is a helicase, involved in DNA repair, and a key regulator of TL.

Lastly, mutations in two additional genes have been reported in DC, although their connection with telomere biology is less straightforward. These include an intronic splice variant in Apollo (encoded by DCLRE1B) in a patient with HH and normal telomeres.[94] The role of Apollo in telomere biology is being investigated. In another study, linkage analysis led to the identification of mutations in C16orf57, a gene with unknown function now called *USB1*.[95] *USB1* mutations were reported not only in patients with DC and normal telomeres but also in individuals with Rothmund–Thomson syndrome and poikiloderma with neutropenia, suggesting an overlapping clinical, but not biological, spectrum.[95]

In total, it is estimated that the germ-line genetic cause of DC is now known in about 70% of families with DC. Ongoing genomic efforts seek to discover the underlying genetic etiology in the remainder of DC/TBD families.

6. DIAGNOSING TBDs

In patients with the classic mucocutaneous triad, the diagnosis of DC is relatively straightforward. In atypical patients, diagnosis is often more challenging. Vulliamy *et al.* proposed the suggested clinical criteria for the diagnosis of DC in 2006. These criteria require (1) the presence of the three features of the mucocutaneous triad (dysplastic fingernails and/or toenails, oral leukoplakia, and lacy, reticular skin pigmentation) or (2) one feature of the triad plus BMF and two other clinical problems seen in DC.[96] These criteria are helpful in diagnosing relatively straightforward cases of DC. However, underrecognition of DC can occur because the triad typically evolves over time and the entire constellation of features may not be appreciated. The genetic heterogeneity of DC further complicates the diagnosis because of the multiple modes of inheritance, nine known associated genes, and variable penetrance and expressivity.

TRF measurement of TL by Southern blots was the first to show that TL in DC patients with *DKC1* mutations is significantly shorter than in healthy controls.[89] Flow FISH TL measurement was shown to be highly sensitive and specific in differentiating patients with DC from their unaffected relatives, from patients with other IBMFS, and from healthy controls.[37,38,97] The initial study of flow FISH TL in DC showed that the presence of telomeres less than the first percentile for age was more than 95% sensitive and specific for differentiating patients with DC from their unaffected relatives and from patients with other IBMFS.[38] The follow-up study confirmed this

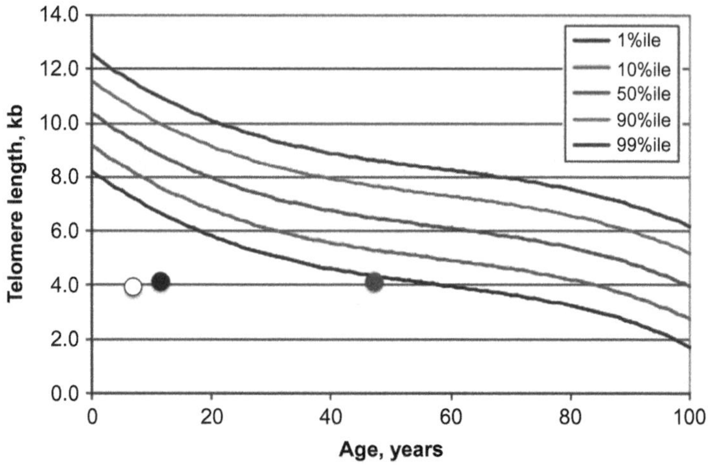

Figure 2.5 *Example of lymphocyte telomere length measured by flow cytometry with fluorescence in situ hybridization (flow FISH).* Open circle, patient with classic dyskeratosis congenita due to a *TINF2* mutation; black circle, patient with dyskeratosis congenita due to a *DKC1* mutation; light gray, patient bone marrow failure, nail dysplasia, and hip avascular necrosis due to a *TERC* mutation. kb, kilobases; %ile, percentile.

observation and also found that patients with *DCK1* and *TINF2* mutations tend to have telomeres that are shorter than those with *TERT* or *TERC* mutations.[37] Patients with more DC-associated clinical complications also had shorter TL than less severely affected individuals (Fig. 2.5).

Measurement of TL by other methods, including qPCR and TR blots, is used in research on DC, but studies of the diagnostic sensitivity and specificity of these methods have not been conducted.[39,40] Currently, TL by flow FISH in leukocyte subsets is the only clinically certified test for DC.

7. GENETIC COUNSELING CONSIDERATIONS

As understanding of the genetic etiology of DC-related TBDs grows, so does the responsibility to provide appropriate genetic education and counseling to the patients and families. Genetic testing for DC-related mutations has implications for the entire family because clinically silent carriers may be identified. The long-term medical complications for these silent carriers are not yet understood and require longitudinal study. All patients and their family members should receive education and counseling that explains the genetic and clinical heterogeneity present in DC and TBDs. They should understand basic concepts of the inheritance of traits and disease.

It is helpful to include information on how DNA is transcribed to RNA and then translated into protein and that the alterations in the protein function are typically responsible for the disease.

For young children, the parents should be educated and counseled. If the child is of the age of assent (usually around 11 or 12 years of age, depending on institutional policies), age-specific education and counseling should be performed with the appropriate assent of the minor child. All individuals undergoing genetic testing should also understand that genetic testing has implications for the entire family. Healthy individuals may be found to be "silent carriers" of a mutation that causes clinically significant disease in their relative.

Couples from families with TBDs may chose to undergo preimplantation genetic diagnosis to attempt to have a healthy, unaffected child.[98] In this instance, the genetic cause of disease in the family needs to be known because the embryos will be tested for the specific mutation in that couple. This is not feasible in mutation-negative families where mutation discovery still needs to occur.

8. CLINICAL MANAGEMENT

Most of the recommendations for clinical management of DC are based on those for Fanconi anemia because there are no current, evidence-based data on medical surveillance strategies in DC.[99,100] The Fanconi anemia guidelines are useful since Fanconi anemia is another rare cancer-prone IBMFS with similar complications, including BMF, cancer, and developmental problems.[100,101] Regardless, each patient must have a clinical management plan tailored to his or her specific needs. Suggestions for clinical surveillance of patients with TBDs are shown in Table 2.2.

8.1. Bone marrow failure

Clinically significant cytopenias are a major problem for patients with DC and related TBDs. In many instances, BMF may be the presenting feature of this spectrum of illnesses. All patients with new-onset BMF should first be evaluated for Fanconi anemia by chromosome breakage analysis. If that test is normal, clinical TL testing should be performed using flow FISH in peripheral blood leukocytes.[37]

Allogeneic HSCT is the only opportunity to cure BMF in DC.[102] Data on the outcome after HSCT in patients with DC are sparse because of the rarity of this disease.[103–111] The largest case series reported, to date, consists

Table 2.2 Clinical surveillance guidelines for individuals with dyskeratosis congenita

Problem	**Suggested surveillance**
Bone marrow failure	Management depends on the severity. If CBCs are normal, consider an annual CBC to identify trends and early manifestations Baseline bone marrow aspiration and biopsy with careful morphological examination and cytogenetic studies. Consider yearly bone marrow evaluation CBCs and bone marrow evaluation should be obtained more frequently if cytopenias are present and at the discretion of the primary hematologist
Bone marrow failure—patients on androgens	Special monitoring is required for patients on androgens for BMF Check liver function tests prior to starting and then every 3 months Perform liver ultrasound examination prior to initiation and semiannually for adenomas, carcinomas, or fibrosis Check cholesterol and triglycerides prior to starting and every 6 months Carefully follow growth and obtain baseline bone age in pediatric patients. Consider endocrinology evaluation
Cancer	Most solid tumors develop after the first decade of life. Patient should be taught how to perform a monthly self-examination for oral, head, and neck cancer Annual cancer screening by a dentist and an otolaryngologist. Follow oral leukoplakia carefully and biopsy any changes or suspicious sites Annual gynecologic evaluation for females Annual dermatologic evaluation
Pulmonary fibrosis	Annual pulmonary function tests are recommended at diagnosis or at an age when the patient can properly perform the test. Early evaluation for shortness of breath or unexplained cough Counsel patients to stop smoking, if applicable
Dental and otolaryngology	Dental hygiene and screening every 6 months Maintain good oral hygiene Inform the primary dentist of the patient's increased risk of oral, head, and neck squamous cell cancers Carefully monitor oral leukoplakia and biopsy suspicious lesions early

Table 2.2 Clinical surveillance guidelines for individuals with dyskeratosis congenita—cont'd

Problem	Suggested surveillance
Ophthalmic	Annual (or more frequent, if needed) exam to detect/correct vision problems, abnormally growing eyelashes, and blocked tear ducts and to look for retinal changes, bleeding, cataracts, and glaucoma
Endocrinology	Baseline bone density scan to evaluate for osteopenia. Follow-up bone density scans yearly or as recommended by physician Careful monitoring of growth and evaluation as needed
Development	Thorough evaluation for developmental delay and therapy/support, as needed
Gastrointestinal and hepatic	Evaluate for clinical history suspicious for esophageal stenosis and/or enteropathy and refer as needed. Obtain baseline liver function tests and use caution when administering potential hepatotoxic medications

These are guidelines only. The primary treatment team should tailor a plan to address each patient's specific need(s).[53]

of 34 patients with DC undergoing HSCT between 1981 and 2009. Unfortunately, the 10-year probability of survival in this group was only 30%. The best outcomes were achieved with sibling HLA-matched HSCT and cyclophosphamide containing nonradiation regimens, but this study was limited by its historical nature and the use of numerous HSCT protocols. Currently, reduced intensity regimens are being studied in order to improve outcomes. HSCT for DC should be performed at centers experienced with this rare disorder.

It is very important to test the relatives of DC patients who are being considered as bone marrow donors prior to HSCT because of the clinical heterogeneity of DC and the presence of silent carriers. There are two case reports of a related HSCT donor being identified as a mutation carrier only after either failure to engraft or failure to mobilize stem cells for their relative with DC. Unfortunately, this resulted in failure of the HSCT and death of the patients.[112,113]

In some instances, patients with DC may be unable to undergo HSCT due to the presence of comorbid conditions, lack of a suitable donor, or personal reasons. Anabolic–androgenic steroids (androgens) have been used to treat acquired and inherited BMF, both prior to the current era of HSCT and currently in patients who are not HSCT candidates.[114,115] A recent case series of 16 DC patients treated with androgens suggested that approximately half of patients with DC had a hematologic response to androgens that led to respond to red blood cell and/or platelet transfusion independence.[116] However, patients on androgens need to be carefully monitored for androgen-related side effects, such as liver function abnormalities, abnormal lipid and cholesterol levels, liver adenomas, and pubertal and growth acceleration. Hematopoietic growth factors may be useful in BMF. Splenic peliosis and splenic rupture were reported in two individuals with DC who received the combination of androgens and granulocyte colony stimulating factor (G-CSF).[117] G-CSF with erythropoietin has occasionally been useful but should not be used in combination with androgens.

8.2. Pulmonary fibrosis

Individuals with TBDs are at a high risk of pulmonary fibrosis, and as described above, pulmonary fibrosis may be the first presenting sign of a DC-related TBD. The optimal management of pulmonary fibrosis due to a TBD is not yet known although clinical trials are underway that may advance understanding. Lung transplantation for pulmonary fibrosis after HSCT has been successfully reported in one patient with DC.[118] Other than lung transplantation, the management of TBD-related pulmonary fibrosis is primarily supportive with pulmonary rehabilitation therapy and the administration of supplemental oxygen.

9. SUMMARY AND FUTURE DIRECTIONS

DC is the prototypical disorder of telomere biology, but a broad spectrum of phenotypes has been uncovered. I have proposed the use of the descriptive term TBD to encompass the broad phenotypic heterogeneity seen due to germ-line mutations in key telomere biology genes. This includes the clinically severe variants of HH and RS, classic DC, and apparently isolated aplastic anemia, pulmonary fibrosis, or other disorders caused by germ-line mutations in genes within the DC/telomere biology pathway. In all of these settings, genetic counseling and education should be provided to the patients. Patients should also be counseled for the potential increased risk of cancer or other TBD-related disorders, such as pulmonary fibrosis. With the availability of clinical TL testing and an increasing awareness of the variable clinical presentations, it is likely that clinicians in many different subspecialties will recognize more patients with TBDs. As laboratory research leads to a greater understanding of telomere biology and the remainder of the genes that can cause TBDs when mutated are discovered, it may become possible to develop directed therapies aimed at improving the quality of life of patients with DC and the related TBDs.

ACKNOWLEDGMENTS

I thank the patients and their families whose valuable contributions have significantly advanced our understanding of telomere biology. This work was supported by the intramural research program of the Division of Cancer Epidemiology and Genetics, National Cancer Institute, National Institutes of Health.

REFERENCES

1. Meyne J, Ratliff RL, Moyzis RK. Conservation of the human telomere sequence (TTAGGG)n among vertebrates. *Proc Natl Acad Sci USA*. 1989;86(18):7049–7053.
2. de Lange T, Shiue L, Myers RM, et al. Structure and variability of human chromosome ends. *Mol Cell Biol*. 1990;10(2):518–527.
3. Wright WE, Tesmer VM, Huffman KE, Levene SD, Shay JW. Normal human chromosomes have long G-rich telomeric overhangs at one end. *Genes Dev*. 1997;11(21):2801–2809.
4. Hockemeyer D, Sfeir AJ, Shay JW, Wright WE, de Lange T. POT1 protects telomeres from a transient DNA damage response and determines how human chromosomes end. *EMBO J*. 2005;24(14):2667–2678.
5. Olovnikov AM. A theory of marginotomy. The incomplete copying of template margin in enzymic synthesis of polynucleotides and biological significance of the phenomenon. *J Theor Biol*. 1973;41(1):181–190.
6. Watson JD. Origin of concatemeric T7 DNA. *Nat New Biol*. 1972;239(94):197–201.
7. Lundblad V, Szostak JW. A mutant with a defect in telomere elongation leads to senescence in yeast. *Cell*. 1989;57(4):633–643.
8. Zou Y, Sfeir A, Gryaznov SM, Shay JW, Wright WE. Does a sentinel or a subset of short telomeres determine replicative senescence? *Mol Biol Cell*. 2004;15(8):3709–3718.
9. Neumann AA, Watson CM, Noble JR, Pickett HA, Tam PP, Reddel RR. Alternative lengthening of telomeres in normal mammalian somatic cells. *Genes Dev*. 2013;27(1):18–23.
10. Shay JW, Wright WE. Role of telomeres and telomerase in cancer. *Semin Cancer Biol*. 2011;21(6):349–353.
11. Greider CW, Blackburn EH. Identification of a specific telomere terminal transferase activity in Tetrahymena extracts. *Cell*. 1985;43(2 Pt 1):405–413.
12. Blackburn EH. Telomeres and telomerase: their mechanisms of action and the effects of altering their functions. *FEBS Lett*. 2005;579(4):859–862.
13. Cohen SB, Graham ME, Lovrecz GO, Bache N, Robinson PJ, Reddel RR. Protein composition of catalytically active human telomerase from immortal cells. *Science*. 2007;315(5820):1850–1853.
14. Hamma T, Ferre-D'Amare AR. The box H/ACA ribonucleoprotein complex: interplay of RNA and protein structures in post-transcriptional RNA modification. *J Biol Chem*. 2010;285(2):805–809.

15. Trahan C, Martel C, Dragon F. Effects of dyskeratosis congenita mutations in dyskerin, NHP2 and NOP10 on assembly of H/ACA pre-RNPs. *Hum Mol Genet.* 2010;19(5):825–836.
16. Zeng XL, Thumati NR, Fleisig HB, et al. The accumulation and not the specific activity of telomerase ribonucleoprotein determines telomere maintenance deficiency in X-linked dyskeratosis congenita. *Hum Mol Genet.* 2012;21(4):721–729.
17. Bellodi C, McMahon M, Contreras A, et al. H/ACA small RNA dysfunctions in disease reveal key roles for noncoding RNA modifications in hematopoietic stem cell differentiation. *Cell Rep.* 2013;3(5):1493–1502.
18. Mitchell JR, Cheng J, Collins K. A box H/ACA small nucleolar RNA-like domain at the human telomerase RNA 3' end. *Mol Cell Biol.* 1999;19(1):567–576.
19. Leulliot N, Godin KS, Hoareau-Aveilla C, et al. The box H/ACA RNP assembly factor Naf1p contains a domain homologous to Gar1p mediating its interaction with Cbf5p. *J Mol Biol.* 2007;371(5):1338–1353.
20. Venteicher AS, Artandi SE. TCAB1: driving telomerase to Cajal bodies. *Cell Cycle.* 2009;8(9):1329–1331.
21. Ulaner GA, Hu JF, Vu TH, Giudice LC, Hoffman AR. Telomerase activity in human development is regulated by human telomerase reverse transcriptase (hTERT) transcription and by alternate splicing of hTERT transcripts. *Cancer Res.* 1998;58(18):4168–4172.
22. Harle-Bachor C, Boukamp P. Telomerase activity in the regenerative basal layer of the epidermis inhuman skin and in immortal and carcinoma-derived skin keratinocytes. *Proc Natl Acad Sci USA.* 1996;93(13):6476–6481.
23. Hiyama K, Hirai Y, Kyoizumi S, et al. Activation of telomerase in human lymphocytes and hematopoietic progenitor cells. *J Immunol.* 1995;155(8):3711–3715.
24. Chiu CP, Dragowska W, Kim NW, et al. Differential expression of telomerase activity in hematopoietic progenitors from adult human bone marrow. *Stem Cells.* 1996;14(2):239–248.
25. Shay JW, Wright WE. Telomerase activity in human cancer. *Curr Opin Oncol.* 1996;8(1):66–71.
26. Conomos D, Pickett HA, Reddel RR. Alternative lengthening of telomeres: remodeling the telomere architecture. *Front Oncol.* 2013;3:27.
27. Dunham MA, Neumann AA, Fasching CL, Reddel RR. Telomere maintenance by recombination in human cells. *Nat Genet.* 2000;26(4):447–450.
28. Aubert G, Hills M, Lansdorp PM. Telomere length measurement-caveats and a critical assessment of the available technologies and tools. *Mutat Res.* 2012;730(1–2):59–67.
29. Baird DM. New developments in telomere length analysis. *Exp Gerontol.* 2005;40(5):363–368.
30. Baird DM, Rowson J, Wynford-Thomas D, Kipling D. Extensive allelic variation and ultrashort telomeres in senescent human cells. *Nat Genet.* 2003;33(2):203–207.
31. Bendix L, Horn PB, Jensen UB, Rubelj I, Kolvraa S. The load of short telomeres, estimated by a new method, Universal STELA, correlates with number of senescent cells. *Aging Cell.* 2010;9(3):383–397.
32. Cawthon RM. Telomere measurement by quantitative PCR. *Nucleic Acids Res.* 2002;30(10):e47.
33. Cawthon RM. Telomere length measurement by a novel monochrome multiplex quantitative PCR method. *Nucleic Acids Res.* 2009;37(3):e21.
34. Elbers CC, Garcia ME, Kimura M, et al. Comparison between southern blots and qPCR analysis of leukocyte telomere length in the health ABC study. *J Gerontol A Biol Sci Med Sci.* 2014;69:527–531.
35. Aviv A, Hunt SC, Lin J, Cao X, Kimura M, Blackburn E. Impartial comparative analysis of measurement of leukocyte telomere length/DNA content by Southern blots and qPCR. *Nucleic Acids Res.* 2011;39(20):e134.

36. Baerlocher GM, Vulto I, de Jong G, Lansdorp PM. Flow cytometry and FISH to measure the average length of telomeres (flow FISH). *Nat Protoc.* 2006;1(5):2365–2376.
37. Alter BP, Rosenberg PS, Giri N, Baerlocher GM, Lansdorp PM, Savage SA. Telomere length is associated with disease severity and declines with age in dyskeratosis congenita. *Haematologica.* 2012;97(3):353–359.
38. Alter BP, Baerlocher GM, Savage SA, et al. Very short telomere length by flow fluorescence in situ hybridization identifies patients with dyskeratosis congenita. *Blood.* 2007;110(5):1439–1447.
39. Du HY, Pumbo E, Ivanovich J, et al. TERC and TERT gene mutations in patients with bone marrow failure and the significance of telomere length measurements. *Blood.* 2009;113(2):309–316.
40. Vulliamy TJ, Kirwan MJ, Beswick R, et al. Differences in disease severity but similar telomere lengths in genetic subgroups of patients with telomerase and shelterin mutations. *PLoS One.* 2011;6(9):e24383.
41. Knight SW, Heiss NS, Vulliamy TJ, et al. X-linked dyskeratosis congenita is predominantly caused by missense mutations in the DKC1 gene. *Am J Hum Genet.* 1999;65(1):50–58.
42. Heiss NS, Knight SW, Vulliamy TJ, et al. X-linked dyskeratosis congenita is caused by mutations in a highly conserved gene with putative nucleolar functions. *Nat Genet.* 1998;19(1):32–38.
43. Yamaguchi H, Baerlocher GM, Lansdorp PM, et al. Mutations of the human telomerase RNA gene (TERC) in aplastic anemia and myelodysplastic syndrome. *Blood.* 2003;102(3):916–918.
44. Yamaguchi H, Calado RT, Ly H, et al. Mutations in TERT, the gene for telomerase reverse transcriptase, in aplastic anemia. *N Engl J Med.* 2005;352(14):1413–1424.
45. Tsakiri KD, Cronkhite JT, Kuan PJ, et al. Adult-onset pulmonary fibrosis caused by mutations in telomerase. *Proc Natl Acad Sci USA.* 2007;104(18):7552–7557.
46. Armanios MY, Chen JJ, Cogan JD, et al. Telomerase mutations in families with idiopathic pulmonary fibrosis. *N Engl J Med.* 2007;356(13):1317–1326.
47. Calado RT, Regal JA, Kleiner DE, et al. A spectrum of severe familial liver disorders associate with telomerase mutations. *PLoS One.* 2009;4(11):e7926.
48. Anderson BH, Kasher PR, Mayer J, et al. Mutations in CTC1, encoding conserved telomere maintenance component 1, cause Coats plus. *Nat Genet.* 2012;44(3):338–342.
49. Polvi A, Linnankivi T, Kivela T, et al. Mutations in CTC1, encoding the CTS telomere maintenance complex component 1, cause cerebroretinal microangiopathy with calcifications and cysts. *Am J Hum Genet.* 2012;90(3):540–549.
50. Dokal I. Dyskeratosis congenita in all its forms. *Br J Haematol.* 2000;110(4):768–779.
51. Schroeder TM, Hofbauer M. Letter: dyskeratosis congenita Zinsser-Cole-Engman form with abnormal karyotype. *Dermatologica.* 1975;151(5):316–318.
52. Dokal I. Dyskeratosis congenita. *Hematology Am Soc Hematol Educ Program.* 2011;2011:480–486.
53. Ballew BJ, Savage SA. Updates on the biology and management of dyskeratosis congenita and related telomere biology disorders. *Expert Rev Hematol.* 2013;6(3):327–337.
54. Walne AJ, Dokal I. Advances in the understanding of dyskeratosis congenita. *Br J Haematol.* 2009;145(2):164–172.
55. Savage SA, Bertuch AA. The genetics and clinical manifestations of telomere biology disorders. *Genet Med.* 2010;12(12):753–764.
56. Dokal I, Vulliamy T, Mason P, Bessler M. Clinical utility gene card for: dyskeratosis congenita. *Eur J Hum Genet.* 2011;19. doi: 10.1038/ejhg.2011.90.
57. Jyonouchi S, Forbes L, Ruchelli E, Sullivan KE. Dyskeratosis congenita: a combined immunodeficiency with broad clinical spectrum—a single-center pediatric experience. *Pediatr Allergy Immunol.* 2011;22(3):313–319.

58. Atkinson JC, Harvey KE, Domingo DL, et al. Oral and dental phenotype of dyskeratosis congenita. *Oral Dis.* 2008;14(5):419–427.
59. Tsilou ET, Giri N, Weinstein S, Mueller C, Savage SA, Alter BP. Ocular and orbital manifestations of the inherited bone marrow failure syndromes: Fanconi anemia and dyskeratosis congenita. *Ophthalmology.* 2010;117(3):615–622.
60. Rackley S, Pao M, Seratti GF, et al. Neuropsychiatric conditions among patients with dyskeratosis congenita: a link with telomere biology? *Psychosomatics.* 2012;53(3):230–235.
61. Alter BP, Giri N, Savage SA, Rosenberg PS. Cancer in dyskeratosis congenita. *Blood.* 2009;113(26):6549–6557.
62. Hoyeraal HM, Lamvik J, Moe PJ. Congenital hypoplastic thrombocytopenia and cerebral malformations in two brothers. *Acta Paediatr Scand.* 1970;59(2):185–191.
63. Hreidarsson S, Kristjansson K, Johannesson G, Johannsson JH. A syndrome of progressive pancytopenia with microcephaly, cerebellar hypoplasia and growth failure. *Acta Paediatr Scand.* 1988;77(5):773–775.
64. Berthet F, Caduff R, Schaad UB, et al. A syndrome of primary combined immunodeficiency with microcephaly, cerebellar hypoplasia, growth failure and progressive pancytopenia. *Eur J Pediatr.* 1994;153(5):333–338.
65. Kirwan M, Dokal I. Dyskeratosis congenita: a genetic disorder of many faces. *Clin Genet.* 2008;73(2):103–112.
66. Ballew BJ, Joseph V, De S, et al. A recessive founder mutation in regulator of telomere elongation helicase 1, RTEL1, underlies severe immunodeficiency and features of Hoyeraal Hreidarsson syndrome. *PLoS Genet.* 2013;9(8):e1003695.
67. Ballew BJ, Yeager M, Jacobs K, et al. Germline mutations of regulator of telomere elongation helicase 1, RTEL1, in dyskeratosis congenita. *Hum Genet.* 2013;132(4):473–480.
68. Savage SA, Giri N, Baerlocher GM, Orr N, Lansdorp PM, Alter BP. TINF2, a component of the shelterin telomere protection complex, is mutated in dyskeratosis congenita. *Am J Hum Genet.* 2008;82(2):501–509.
69. Walne AJ, Vulliamy T, Beswick R, Kirwan M, Dokal I. TINF2 mutations result in very short telomeres: analysis of a large cohort of patients with dyskeratosis congenita and related bone marrow failure syndromes. *Blood.* 2008;112(9):3594–3600.
70. Walne AJ, Vulliamy T, Marrone A, et al. Genetic heterogeneity in autosomal recessive dyskeratosis congenita with one subtype due to mutations in the telomerase-associated protein NOP10. *Hum Mol Genet.* 2007;16(13):1619–1629.
71. Vulliamy T, Beswick R, Kirwan M, et al. Mutations in the telomerase component NHP2 cause the premature ageing syndrome dyskeratosis congenita. *Proc Natl Acad Sci USA.* 2008;105(23):8073–8078.
72. Revesz T, Fletcher S, al-Gazali LI, DeBuse P. Bilateral retinopathy, aplastic anaemia, and central nervous system abnormalities: a new syndrome? *J Med Genet.* 1992;29(9):673–675.
73. Savage SA. Connecting complex disorders through biology. *Nat Genet.* 2012;44(3):238–240.
74. Scheinberg P, Cooper JN, Sloand EM, Wu CO, Calado RT, Young NS. Association of telomere length of peripheral blood leukocytes with hematopoietic relapse, malignant transformation, and survival in severe aplastic anemia. *JAMA.* 2010;304(12):1358–1364.
75. Cronkhite JT, Xing C, Raghu G, et al. Telomere shortening in familial and sporadic pulmonary fibrosis. *Am J Respir Crit Care Med.* 2008;178(7):729–737.
76. Borie R, Crestani B, Bichat H. Prevalence of telomere shortening in familial and sporadic pulmonary fibrosis is increased in men. *Am J Respir Crit Care Med.* 2009;179(11):1073.
77. Calado RT, Brudno J, Mehta P, et al. Constitutional telomerase mutations are genetic risk factors for cirrhosis. *Hepatology.* 2011;53(5):1600–1607.
78. Goldman F, Bouarich R, Kulkarni S, et al. The effect of TERC haploinsufficiency on the inheritance of telomere length. *Proc Natl Acad Sci USA.* 2005;102(47):17119–17124.

79. Armanios M, Chen JL, Chang YP, et al. Haploinsufficiency of telomerase reverse transcriptase leads to anticipation in autosomal dominant dyskeratosis congenita. *Proc Natl Acad Sci USA*. 2005;102(44):15960–15964.
80. Vulliamy T, Marrone A, Szydlo R, Walne A, Mason PJ, Dokal I. Disease anticipation is associated with progressive telomere shortening in families with dyskeratosis congenita due to mutations in TERC. *Nat Genet*. 2004;36(5):447–449.
81. Savage SA. Dyskeratosis congenita. In: Pagon RA, Adam MP, Ardinger HH, et al. *GeneReviews*. Seattle, WA: University of Washington, Seattle; 1993.
82. Vulliamy T, Marrone A, Goldman F, et al. The RNA component of telomerase is mutated in autosomal dominant dyskeratosis congenita. *Nature*. 2001;413(6854):432–435.
83. Vulliamy TJ, Walne A, Baskaradas A, Mason PJ, Marrone A, Dokal I. Mutations in the reverse transcriptase component of telomerase (TERT) in patients with bone marrow failure. *Blood Cells Mol Dis*. 2005;34(3):257–263.
84. Zhong F, Savage SA, Shkreli M, et al. Disruption of telomerase trafficking by TCAB1 mutation causes dyskeratosis congenita. *Genes Dev*. 2011;25(1):11–16.
85. Keller RB, Gagne KE, Usmani GN, et al. CTC1 mutations in a patient with dyskeratosis congenita. *Pediatr Blood Cancer*. 2012;59(2):311–314.
86. Walne AJ, Vulliamy T, Kirwan M, Plagnol V, Dokal I. Constitutional mutations in RTEL1 cause severe dyskeratosis congenita. *Am J Hum Genet*. 2013;92(3):448–453.
87. Le Guen T, Jullien L, Touzot F, et al. Human RTEL1 deficiency causes Hoyeraal-Hreidarsson syndrome with short telomeres and genome instability. *Hum Mol Genet*. 2013;22:3239–3249.
88. Knight SW, Heiss NS, Vulliamy TJ, et al. Unexplained aplastic anaemia, immunodeficiency, and cerebellar hypoplasia (Hoyeraal-Hreidarsson syndrome) due to mutations in the dyskeratosis congenita gene, DKC1. *Br J Haematol*. 1999;107(2):335–339.
89. Mitchell JR, Wood E, Collins K. A telomerase component is defective in the human disease dyskeratosis congenita. *Nature*. 1999;402(6761):551–555.
90. Diaz de Leon A, Cronkhite JT, Katzenstein AL, et al. Telomere lengths, pulmonary fibrosis and telomerase (TERT) mutations. *PLoS One*. 2010;5(5):e10680.
91. Armanios MY, Chen JJ, Cogan JD, et al. Telomerase mutations in families with idiopathic pulmonary fibrosis. *N Eng J Med*. 2007;356(13):1317–1326.
92. Du HY, Mason PJ, Bessler M, Wilson DB. TINF2 mutations in children with severe aplastic anemia. *Pediatr Blood Cancer*. 2009;52(5):687.
93. Walne AJ, Bhagat T, Kirwan M, et al. Mutations in the telomere capping complex in bone marrow failure and related syndromes. *Haematologica*. 2013;98(3):334–338.
94. Touzot F, Callebaut I, Soulier J, et al. Function of Apollo (SNM1B) at telomere highlighted by a splice variant identified in a patient with Hoyeraal-Hreidarsson syndrome. *Proc Natl Acad Sci USA*. 2010;107(22):10097–10102.
95. Walne AJ, Vulliamy T, Beswick R, Kirwan M, Dokal I. Mutations in C16orf57 and normal-length telomeres unify a subset of patients with dyskeratosis congenita, poikiloderma with neutropenia and Rothmund-Thomson syndrome. *Hum Mol Genet* 2010;19(22):4453–4461.
96. Vulliamy TJ, Marrone A, Knight SW, Walne A, Mason PJ, Dokal I. Mutations in dyskeratosis congenita: their impact on telomere length and the diversity of clinical presentation. *Blood*. 2006;107(7):2680–2685.
97. Gadalla SM, Cawthon R, Giri N, Alter BP, Savage SA. Telomere length in blood, buccal cells, and fibroblasts from patients with inherited bone marrow failure syndromes. *Aging (Albany, NY)*. 2010;2(11):867–874.
98. Kuliev A, Rechitsky S, Tur-Kaspa I, Verlinsky Y. Preimplantation genetics: improving access to stem cell therapy. *Ann N Y Acad Sci*. 2005;1054:223–227.
99. Savage SA, Dokal I, Armanios M, et al. Dyskeratosis congenita: the first NIH clinical research workshop. *Pediatr Blood Cancer*. 2009;53(3):520–523.

100. Eiler ME, Frohnmayer L, Larsen K, Owen J, eds. *Fanconi Anemia: Guidelines for Diagnosis and Management.* 3rd ed. Eugene, OR: Fanconi Anemia Research Fund, Inc.; 2008.
101. Shimamura A, Alter BP. Pathophysiology and management of inherited bone marrow failure syndromes. *Blood Rev.* 2010;24(3):101–122.
102. Al-Rahawan MM, Giri N, Alter BP. Intensive immunosuppression therapy for aplastic anemia associated with dyskeratosis congenita. *Int J Hematol.* 2006;83(3):275–276.
103. Berthou C, Devergie A, D'Agay MF, et al. Late vascular complications after bone marrow transplantation for dyskeratosis congenita. *Br J Haematol.* 1991;79(2):335–336.
104. de la Fuente J, Dokal I. Dyskeratosis congenita: advances in the understanding of the telomerase defect and the role of stem cell transplantation. *Pediatr Transplant.* 2007;11(6):584–594.
105. Yabe M, Yabe H, Hattori K, et al. Fatal interstitial pulmonary disease in a patient with dyskeratosis congenita after allogeneic bone marrow transplantation. *Bone Marrow Transplant.* 1997;19(4):389–392.
106. Dror Y, Freedman MH, Leaker M, et al. Low-intensity hematopoietic stem-cell transplantation across human leucocyte antigen barriers in dyskeratosis congenita. *Bone Marrow Transplant.* 2003;31(10):847–850.
107. Brazzola P, Duval M, Fournet JC, et al. Fatal diffuse capillaritis after hematopoietic stem-cell transplantation for dyskeratosis congenita despite low-intensity conditioning regimen. *Bone Marrow Transplant.* 2005;36(12):1103–1105, author reply 1105.
108. Ostronoff F, Ostronoff M, Calixto R, et al. Fludarabine, cyclophosphamide, and antithymocyte globulin for a patient with dyskeratosis congenita and severe bone marrow failure. *Biol Blood Marrow Transplant.* 2007;13(3):366–368.
109. Vuong LG, Hemmati PG, Neuburger S, et al. Reduced-intensity conditioning using fludarabine and antithymocyte globulin alone allows stable engraftment in a patient with dyskeratosis congenita. *Acta Haematol.* 2010;124(4):200–203.
110. Dietz AC, Orchard PJ, Baker KS, et al. Disease-specific hematopoietic cell transplantation: nonmyeloablative conditioning regimen for dyskeratosis congenita. *Bone Marrow Transplant.* 2011;46(1):98–104.
111. Nishio N, Takahashi Y, Ohashi H, et al. Reduced-intensity conditioning for alternative donor hematopoietic stem cell transplantation in patients with dyskeratosis congenita. *Pediatr Transplant.* 2011;15(2):161–166.
112. Fogarty PF, Yamaguchi H, Wiestner A, et al. Late presentation of dyskeratosis congenita as apparently acquired aplastic anaemia due to mutations in telomerase RNA. *Lancet.* 2003;362(9396):1628–1630.
113. Denny CC, Wilfond BS, Peters JA, Giri N, Alter BP. All in the family: disclosure of "unwanted" information to an adolescent to benefit a relative. *Am J Med Genet A.* 2008;146A(21):2719–2724.
114. Diamond LK, Shahidi NT. Treatment of aplastic anemia in children. *Semin Hematol.* 1967;4(3):278–288.
115. Shahidi NT. A review of the chemistry, biological action, and clinical applications of anabolic-androgenic steroids. *Clin Ther.* 2001;23(9):1355–1390.
116. Khincha P, Wentzensen I, Giri N, Alter BP, Savage SA. Response to androgen therapy and side effects in patients with dyskeratosis congenita. *Br J Haematol.* 2014;165:349–357.
117. Giri N, Pitel PA, Green D, Alter BP. Splenic peliosis and rupture in patients with dyskeratosis congenita on androgens and granulocyte colony-stimulating factor. *Br J Haematol.* 2007;138(6):815–817.
118. Giri N, Lee R, Faro A, et al. Lung transplantation for pulmonary fibrosis in dyskeratosis congenita: case Report and systematic literature review. *BMC Blood Disord.* 2011;11:3.

CHAPTER THREE

Telomere Biology in Stem Cells and Reprogramming

Luis F.Z. Batista[*,†]
[*]Department of Medicine, Washington University School of Medicine, St. Louis, Missouri, USA
[†]Department of Developmental Biology, Washington University School of Medicine, St. Louis, Missouri, USA

Contents

Abstract

Telomerase expression in humans is restricted to different populations of stem and progenitor cells, being silenced in most somatic tissues. Efficient telomere homeostasis is essential for embryonic and adult stem cell function and therefore essential for tissue homeostasis throughout organismal life. Accordingly, the mutations in telomerase culminate in reduced stem cell function both *in vivo* and *in vitro* and have been associated with tissue dysfunction in human patients. Despite the importance of telomerase for stem cell biology, the mechanisms behind telomerase regulation during development are still poorly understood, mostly due to difficulties in acquiring and maintaining pluripotent stem cell populations in culture. In this chapter, we will analyze recent developments in this field, including the importance of efficient telomere homeostasis in different stem cell types and the role of telomerase in different techniques used for cellular reprogramming.

Progress in Molecular Biology and Translational Science, Volume 125
ISSN 1877-1173
http://dx.doi.org/10.1016/B978-0-12-397898-1.00003-7

1. INTRODUCTION

1.1. A variety of stem cells throughout organismal development

Stem cells are functionally defined by their ability to proliferate in the same state (self-renew) while being able to generate differentiated cell types in the developing and adult organism.[1] Typically, stem cells are classified by their developmental potential, ranging from totipotent, represented by cells that are able to generate the whole organism, including the extraembryonic trophoblast lineage (zygote and early blastomeres in mammals), to unipotent cells that are able to differentiate into just one particular cell type (such as spermatogonial stem cells). Mouse embryonic stem (ES) cells are derived from the inner cell mass (ICM) of the embryo and represent the pluripotent stem cell state, able to generate the entire organism except the trophoblast lineage.[2,3] In addition, self-renewing, pluripotent stem cells have been derived from the mouse embryo at a slightly later stage, when the ICM has become the epiblast (these cells are then called epiblast stem cells). However, due to ethical concerns and technical challenges, it was not until almost 20 years later that James Thomson and colleagues were able to successfully derive human ES cells from the ICM mass of human blastocysts.[4] The relative ease of culturing mouse and human ES cells *in vitro*, together with the promise that these cells hold for regenerative medicine, prompted an extensive molecular and biochemical characterization of the pluripotent state and provided several insights into organismal development (for detailed reviews on ES cell biology, please see Refs. 5–8). On the other hand, adult stem cells have proven to be harder to isolate and to study *in vitro*.[9] These cells are more restricted in their developmental potential than ES cells and are typically classified as multipotent cells, able to differentiate into all cell types of a particular cellular lineage, being responsible for the replenishing of specific tissues throughout organismal life. The study of the molecular control behind each of these different stem cell states is fundamental for the understanding of mammalian development and human disease. In fact, a growing body of evidence suggests that the number of stem cells in adult tissues must be under strict genetic control in order to avoid uncontrolled expansion or exhaustion of a particular tissue, which can lead to different human conditions, such as cancer and aging.[10,11] Efficient stem cell function is therefore essential for organismal fitness. In this chapter, we will focus on how telomerase activity and telomere homeostasis are controlled in different stem cell populations and on the importance of this process for proper stem cell self-renewal, a process that ultimately maintains the stem cell pool for the life of an organism.

1.2. Telomeres cap chromosomal ends

Telomeres are the physical ends of eukaryotic chromosomes. In vertebrates, these sequences are composed of long stretches of TTAGGG repeats that can extend up to 15 kb in humans and 100 kb in rodents. The telomeric DNA is composed of a long double-stranded tract that ends in a short, single-stranded overhang. This single-stranded unit of the telomere invades the double-stranded sequence, giving rise to a lasso-like structure named "T-loop" that protects the telomere terminus from the DNA damage-repair machinery that would otherwise recognize telomeres as potential DNA double-strand breaks.[12] Providing further protection to chromosome ends, telomeres are also bound by shelterin, a large multisubunit protein complex that prevents the chromosome ends from being recognized as a DNA break and inhibits inappropriate recombination. The physiological relevance of efficient telomere protection is well demonstrated by the inhibition or deletion of specific shelterin components, where there is rapid telomere uncapping, which results in a local DNA damage response at chromosome ends, leading to robust activation of DNA damage pathways.[13–17] More recently, it was discovered that in addition to shelterin, telomeres in vertebrates are also bound by proteins belonging to the CST complex, indicating that proper telomere homeostasis might result from cooperation between shelterin and CST to keep the 3′ overhang telomeric DNA stable.[18,19]

2. TELOMERE HOMEOSTASIS IN STEM CELLS

2.1. Telomerase activity in pluripotent stem cells

Due to the inability of DNA polymerases to fully replicate chromosome ends at the lagging strand[20] at every cell division, there is a loss of up to 200 bp of telomeric DNA in mammalian cells.[21] This is especially relevant for ES cells, since these have a rapid cell cycle progression characterized by an abbreviated G1 phase of the cell cycle.[22,23] This fast cell cycle is a defining characteristic of pluripotent cells, and it is thought to be necessary for continual self-renewal and to resist differentiation,[24] where the forced expression genes that regulate progression through the cell cycle induce fast and potent cellular differentiation.[25,26] To avoid the significant telomere shortening that would otherwise be associated with continuous cellular divisions, mammalian pluripotent cells have high activity levels of telomerase, the ribonucleoprotein (RNP) enzyme responsible for synthesizing telomeres.[27–29] The importance of telomerase activity for the homeostasis and proliferation of pluripotent cells is clear from experiments using telomerase-ablated ES cells, which showed progressive telomere shortening, genomic instability, aneuploidy, and telomere fusions that culminated with reduced growth rates.[30] The only cells that maintained proliferation were the ones that were able to elongate their telomeres using telomerase-independent mechanisms.[31–33] Interestingly, it has recently been shown that mouse ES cells show a unique mode of telomere maintenance that relies on the transient expression of ZSCAN4, which promotes a rapid telomere extension by telomere recombination.[34] Interestingly, while the developmental potency of mouse ES cells is known to deteriorate during long-term cell culture, increasing the frequency of Zscan4 activation in mouse ES cells restores and maintains their developmental potency in long-term cell culture.[35] Confirming the importance of telomerase for pluripotency potential, tetraploid embryo complementation experiments (the most stringent test for pluripotency) have shown that telomerase-deficient mESCs with short telomeres lose their ability to generate complete ESC pups.[36] On the other hand, the overexpression of telomerase enhanced self-renewal, improved resistance to apoptosis, and increased proliferation in mouse[37] and human[38] ES cells. Biochemically, telomerase is a large, multisubunit protein complex that, in addition to TERT, its reverse transcriptase component is composed of TERC, the telomerase RNA component used as a template for the transcriptase reaction; dyskerin (DKC1), a TERC-binding protein necessary for its stability; and TCAB1, necessary for telomerase translocation to Cajal bodies and holoenzyme assembly correction.[39–41] While TERC is expressed in several somatic tissues, TERT is only found in stem and progenitor cells, rendering the vast majority of human adult somatic cells telomerase negative.[42] The high levels of TERT in pluripotent cells culminate in telomere elongation during the derivation of both mouse[43] and human[44] ES cells from the ICM of blastocysts. In addition, both TERT expression and telomerase activity are rapidly downregulated during differentiation of pluripotent cells,[45,46] which indicates that they can be regarded as markers of undifferentiated ES cell populations.

2.2. Telomerase activity in adult multipotent cells

Most adult somatic tissues present a complex cellular hierarchy that starts with a small number of multipotent stem cells that can either self-renew (thereby maintaining the stem cell pool) or differentiate into progenitor cells that have a limited proliferative potential and usually differentiate into mature, functional, tissue somatic cells.[47] Adult stem cells are normally detected as long-lived cellular populations that reside in specialized niches of a given tissue and are maintained in a quiescent state. However, a growing body of evidence indicates the existence of long-lived, yet actively cycling, adult stem cell populations.[48–50] These cells would support the homeostasis of rapidly dividing organs, such as the gut and the skin, and the hematopoietic system. To be able to maintain self-renewal, while actively cycling to generate progenitor cells, these active adult stem cell compartments must retain efficient telomere maintenance mechanisms. Accordingly, despite difficulties in isolating and characterizing adult stem cells, telomerase activity has been detected in different adult stem cell populations. We will now analyze different adult stem cell compartments in more detail.

2.2.1 Hematopoietic system

Samples from the hematopoietic system are more directly obtainable than most other adult tissues, which translated to more detailed research on the functional and molecular characterization of its components. Telomerase activity was described in early experiments with the hematopoietic system, including in hematopoietic stem cells that give rise to all other cells in the blood.[51,52] In fact, elegant experiments from the Weissman group demonstrated that the frequency of telomerase-expressing cells within different populations of the hematopoietic system was proportional to the frequency of cells thought to have self-renewal potential. Among bone marrow hematopoietic stem cells, 70% exhibited detectable telomerase activity.[51] However, telomeres in hematopoietic cells, including hematopoietic stem cells (HSCs), shorten during organismal aging in both mice[53] and humans,[54] which indicates that the amount of telomerase activity is insufficient to cope with the constant renewal that occurs in the blood system. The importance of telomerase for the long-term replicative capacity of HSCs was first exemplified in serial transplantation experiments, where telomerase-deficient HSCs could be serially transplanted for only two rounds, whereas wild-type HSCs could be serially transplanted for at least four rounds. During these experiments, the rate of telomere shortening was increased approximately twofold during serial transplantation of telomerase-deficient HSCs.[55] Interestingly, HSCs overexpressing telomerase did not show an increased repopulation capacity, showing that telomere-independent barriers may act independently and limit the transplantation capacity of HSCs.[56]

2.2.2 Hair follicle stem cells

Telomerase activity has also been identified in mitotically active segments of the hair follicle, which later were discovered to be the niche for hair follicle stem cells.[57] Interestingly, telomerase activity was shown to be higher in ultraviolet light-damaged skin, which can indicate either that telomerase is involved in the early stage of skin carcinogenesis or that telomerase helps tissue regeneration.[58] In fact, using telomere shortening inhibited the mobilization of stem cells out of their niche, impaired hair growth, and resulted in the suppression of the regenerative capacity of these cells *in vitro*.[59] On the other hand, induction of TERT in mouse skin epithelium caused a rapid activation of the hair follicle cycle (from telogen, the resting phase, to anagen, the active phase) culminating in robust hair growth.[60]

2.2.3 Intestinal stem cells

The constant mechanical damage inflicted on the gut by passing bowel content, combined with the chemical and biological assault by the luminal contents, leads to a nearly complete renewal of the intestinal epithelium every 4–5 days, therefore constituting the fastest self-renewing tissue of the human body.[61,62] This fantastic rate of self-renewal, together with a very distinct organizational anatomy, transformed the intestine into one of the most studied models in the adult stem cell field. In a tissue with such a constant need for renewal, it is not surprising that telomerase activity was identified in intestinal stem cells.[63–65] Interestingly, while the Breault group used a GFP-TERT transgenic mouse-model approach to conclude that TERT-positive stem cells in the gut are Bmi1-positive and LGR5-negative and represent a rare and slow-cycling stem cell population,[63] Hans Clevers group used a more sensitive technique, where they would sort LGR5-positive cells and analyze telomerase activity by TRAP (telomere repeat amplification protocol) to conclude that the fast-cycling LGR5-positive cells in the gut actually have high telomerase activity.[65,66]

2.2.4 Muscle stem cells

Recently, investigators from Stanford University were able to demonstrate the importance of telomerase activity also for muscle homeostasis. Working with a Duchenne muscular dystrophy mouse model (mdx) lacking the RNA component of telomerase (TERC), the authors show that telomere attrition causes severe muscular dystrophy that progressively worsens with age.[67] The worsening of symptoms was due to the loss of adult muscle stem cells and could be ameliorated histologically by the transplantation of wild-type muscle stem cells in telomerase-deficient mice. These results not only demonstrate the importance of telomere in muscle stem cell homeostasis but also indicate that Duchenne muscular dystrophy is a stem cell disease. Moreover, it was also demonstrated that telomere dysfunction plays a role in cardiac failure in Duchenne muscular dystrophy, where demands of contraction in the absence of dystrophin, coupled with increased oxidative stress in these settings, cause accelerated telomere erosion, which culminates in cardiac failure and death.[68]

2.2.5 Neural stem cells

The identification of neural stem cells (NSCs) has, historically, been challenging. However, several breakthroughs during the last years collectively indicated that the subventricular zone (SVZ) and the subgranular zone of the hippocampus represent neurogenic niches or local microenvironments that permit and support neurogenesis.[69] Because telomerase expression is so closely related to stem cell function, different groups have tried to identify telomerase-positive cells in the brain. Using *sox2* as a prospective marker for NSCs, it was recently shown that sox2-positive cells isolated from the SVZ were capable of sustained mitotic expansion and showed high telomerase activity.[70] Additionally, it was also shown that telomere attrition impaired the proliferation of adult NSCs that were isolated from the SVZ of telomerase-ablated mice. Curiously, the same authors reported that telomere attrition did not affect the *in vivo* proliferation potential of embryonic NSCs.[71] Thus, despite technical limitations that still complicate the study of adult neurogenesis, it seems reasonable to assume that at least, a percentage of NSCs are telomerase-positive and that telomere homeostasis is important for the maintenance of the adult stem cell pool in the brain. Certainly, with higher definition (deeper and to a higher resolution) of *in vivo* imaging techniques, we will be able to track telomerase-positive cells and understand their precise role in neurogenesis.

2.2.6 Spermatogonial stem cells

Throughout an individual's life, the spermatogenic process relies on proper regulation of self-renewal and differentiation of the spermatogonial stem cells. These are rare single cells that comprise about only 0.03% of the total number of germ cells and are situated on the basal membrane of the seminiferous epithelium. Interestingly, telomere length seems to elongate over the human life span, and this is presumed to be due to high telomerase activity in spermatogonial stem cells throughout the life of a male.[72] In fact, although the knowledge about the identity and characteristics of spermatogonial stem cells in human is very limited, initial experiments with adult testes from Rhesus monkeys identified a population, localized at the basement membrane of seminiferous tubules, that showed high telomerase activity.[73] These results have recently been corroborated in spermatogonial stem cells from adult human testis.[74] The presence of telomerase in the human male germ line seems to be functionally relevant, since the presence of single-nucleotide polymorphisms in TERT and in TEP1 (telomerase-associated protein 1) was recently associated with susceptibility to male infertility.[75]

Interestingly, telomerase activity is low or even absent in oocytes and cleavage-stage embryos, and it is only recovered in blastocysts.[42] Accordingly, telomere length in the oocyte is shorter than in somatic cells. The mechanisms responsible for telomere resetting during the early stages of development are fascinating. Telomeres lengthen considerably following activation of the egg (in mice, thousands of base pairs are added within the first two cell cycles). This happens despite low levels of telomerase activity and seems to be driven by extensive telomere sister-chromatid exchange, and accordingly, DNA recombination proteins were colocalized to telomeres in early cleavage-stage embryos.[76] Interestingly, following this early stage and just as telomerase activity increases in blastocysts, the proteins for recombination and DNA damage repair, as well as Telomere-Sister Chromatid Exchange (T-SCE), decrease markedly, suggesting that from this stage on, telomerase is responsible for telomere elongation, which is corroborated by experiments showing that telomere elongation at this stage of differentiation is abrogated in telomerase-deficient mice.[77] This fascinating dual mode of telomere elongation is most likely responsible for the telomere length resetting during the vertical transfer of genomic material through time.

2.3. Organismal consequences of telomere attrition in stem cells

The importance of telomerase activity in organismal fitness was first delineated from the generation and characterization of laboratory mice null for either TERT or TERC. Although early generation (G1) of these mice has a lack of phenotype, late generations (>G3) present short telomeres and an associated shortened life span, decreased fertility, and impaired organ function[78,79] that are more readily observable in tissues with high turnover rates.[80] More recently, reduced tissue renewal capability in high-turnover tissues was also observed in mouse models with mutant shelterin, providing a clear link between telomere dysfunction and tissue renewal in rodents.[81–84] In humans, the importance of proper telomerase function in adult stem cell compartments is readily observable in patients suffering from different syndromes associated with dysfunctional telomeres, such as the inherited bone marrow failure syndrome dyskeratosis congenita (DC).[85,86] To date, all known mutations in DC patients have been found in genes responsible for telomere homeostasis, and an underlying characteristic of this disease is the presence of short telomeres, usually below the first percentile when compared with the rest of the population.[87] Patients with DC present with systemic tissue defects that are more pronounced in organs with high turnover rates, suggesting an exhaustion of actively cycling tissue stem cells due to a telomerase deficiency.[88] More details of human diseases associated with telomere dysfunction are described elsewhere in this book.

2.4. Noncanonical roles of telomerase in stem cell regulation

The experiments described above with germ-line telomerase knockout mice indicate that the deleterious effects of telomerase deficiency are only observed after a long time frame, with the continuous inheritance of progressively shortened telomeres from one generation to the other. However, recent evidence suggests that TERT is able to directly stimulate stem cells, in a mechanism independent from its telomere elongation function. Initial experiments using TERC knockout and TERT catalytically inactive mice showed that conditional induction of TERT in skin epithelium causes rapid activation of hair follicle stem cells present in the bulge region, inducing robust hair growth.[60] This noncanonical role of telomerase in stem cell regulation has been confirmed by different groups, both in mouse[59] and in zebra fish.[89] Analysis of genome-wide transcriptional response to acute changes in TERT levels in mouse skin revealed that the TERT transcriptional response closely correlates with the response mediated by both Myc and Wnt, two proteins that are intimately associated with stem cell function.[90] Corroborating this idea, it has been shown that the TERT noncanonical activity in stem cell activation is mediated by its interaction with Brg1 and β-catenin, the central activator of the canonical Wnt pathway.[91] Interestingly, it was also shown that β-catenin regulates TERT expression through an interaction with Klf4, a major component of the pluripotency transcriptional network.[92] Although the association between TERT and Wnt hasn't always been confirmed,[93,94] the role of TERT as a transcriptional modulator of the Wnt/beta-catenin signaling pathway and its importance in tissue homeostasis have been recently confirmed by Shkreli *et al.* who found that the conditional overexpression of TERT in mouse kidney induces Wnt signaling in podocytes, resulting in a strong proliferative response in these cells.[95]

In addition to its telomere-independent role on Wnt pathway activation, TERT has recently been shown to interact not only with TERC but also with the RNA component of mitochondrial RNA-processing endoribonuclease (RMRP), forming a distinct RNP complex that shows RNA-dependent RNA polymerase activity.[96] The TERT–RMRP ribonucleoprotein has the ability to produce double-stranded RNAs that are subsequently processed by DICER into small interfering RNAs. RMRP is highly expressed in a wide range of human and murine tissues, and mutations in RMRP are found in patients suffering from cartilage–hair hypoplasia, a disease that affects stem cells.[97,98] Therefore, it is possible that the TERT–RMRP complex is also involved in stem cell regulation. Indeed, although this possible role of TERT–RMRP in stem cell regulation is yet to be described, it has recently been shown that RMRP is essential for early murine development.[99]

3. CELLULAR REPROGRAMMING

3.1. Somatic cell nuclear transfer and cell fusion

Cells that become terminally differentiated will usually lose their ability to generate other cell types. However, experiments carried out during the early 1950s provided initial evidence that cellular differentiation could actually be reversible. Using amphibians as a model, Briggs and King performed the first "cloning" experiments reported in the literature, using a technology still in use today, called somatic cell nuclear transfer (SCNT).[100] In these experiments, the authors showed that the transfer of a cell nucleus from an early-blastocyst-stage embryo into an egg cell whose nucleus had been removed resulted in swimming tadpole clones. The now Nobel laureate John Gurdon followed up on those experiments and demonstrated that normal adult frogs could be generated by transferring nuclei from highly specialized tadpole intestinal cells into irradiated egg cells.[101,102] Basically, the nuclear reprogramming using SCNT represents the insertion of a nucleus of a differentiated somatic cell into an enucleated, unfertilized egg cell of the same species. When successful, the differentiated nucleus is reprogrammed back to a zygotic state and has the potential to generate an entire organism that is genetically identical to the donor of the somatic cell. Technical challenges resulted in a long waiting time before the first mammal was cloned. Finally, in 1997, the SCNT strategy was successfully used to fuse an adult sheep mammary cell with an enucleated sheep egg cell, which was then implanted into an adult recipient female sheep.[103] Following that seminal discovery, several other mammals have been successfully cloned using SCNT, including mice, cattle, pigs, rabbits, and cats.[104] In particular, mice have been cloned using similar SCNT strategies, using the nuclei of highly differentiated cells, such as neurons[105] and lymphocytes.[106]

Another strategy used for nuclear reprogramming is cell fusion, which involves fusing of two or more cell types to form a single entity. Cell-fusion studies provided the first evidence that differentiation of mammalian somatic cells is reversible and that it requires continuous regulation.[107–109] More recent experiments using cell-fusion techniques showed that the expression of previously silent genes typical of diverse differentiated mammalian cell types can be induced in other terminally differentiated cell types, even *in vivo*.[110,111] For a detailed review on SCNT and cell-fusion experiments, please see Ref. 112.

3.1.1 Telomere biology in SCNT

A question that immediately arose when the first mammal (the famous Dolly, the sheep) was successfully cloned[103] was if its cells would maintain the telomere length from the adult donor cells or if telomeres would too be reprogrammed to acquire the length of telomeres in ES cells during the process. Frustratingly enough, when telomeres from Dolly where compared with age-matched controls, they were found to be approximately 20% shorter. Likewise, telomeres from other cloned sheep at that time were also shown to be shorter than age-matched controls.[113] This was a clear indication that telomeres were not being reprogrammed during SCNT and could represent a potential barrier to the therapeutic potential of this technique.

However, subsequent studies analyzing telomere lengths in several different cloned animals (using the SCNT technique) showed different and more promising results, with the cloned animals carrying telomeres of similar length of age-matched controls. This was true for mice,[114] cattle,[115–118] and sheep derived from fibroblast cell lines.[119] It is still unclear why the original SCNT experiments with Dolly and other sheep generated offspring with short telomeres. It is also unclear why Dolly died at an early age. Most likely, these differences are the consequence of improved protocols for SCNT cloning and probably intrinsic characteristics of donor cells. Nevertheless, it is now well accepted that telomerase is activated in cloned embryos that were derived from donor nuclei with undetectable levels of telomerase. Interestingly, also in cloned embryos, the activation of telomerase is again found at the blastocyst stage of embryogenesis.[116,117,120] Importantly, a recent study suggests that the telomere elongation in cloned embryos is independent of the type and telomere length of the donor cell and is intrinsic to the cloning process itself.[121]

3.2. Reprogramming through induced pluripotency

The successful cellular dedifferentiation using SCNT and cell-fusion strategies provided proof that mammalian development does not require irreversible changes in the genome. However, those techniques are cumbersome, and SCNT is controversial since it involves the use and destruction of human embryos. Therefore, it is not surprising that the discovery of an easy and fast method that is free of potential ethical issues for cellular reprogramming was received with great exhilaration by the regenerative medicine community. Decades of research in cellular reprogramming have recently culminated in the development of "induced pluripotent stem cells" (iPSCs).[122–124] With this technology, adult somatic cells can be reprogrammed back to a pluripotent state, by forced expression of transcription factors associated with the stem cell state. The most commonly used factors, Oct-4, Sox-2, Klf4, and c-Myc, have been used to reprogram cells from a growing number of species, using a variety of reprogramming methods, including RNA,[125] microRNA,[126] and protein transduction.[127] These cells have represented the biggest breakthrough in regenerative medicine in the last decades, and several groups are actively generating patient-specific iPS cells from a variety of diseases. The hope is that these cells will help in the understanding of developmental aspects of different syndromes and that they can be used as platforms for drug discovery and future transplantation sources (for a detailed review on the use of iPS cells for regenerative medicine, see Ref. 128). However, before these cells can be used in clinical settings, we need to confirm that they represent, in fact, a pluripotent state that is safe for human experimentation. Since telomerase activity and telomere attrition are so fundamental for stem cell biology, it comes as no surprise that a lot of emphases are given to understanding telomerase and telomere regulation in settings of induced reprogramming by defined factors.

3.2.1 Telomere reprogramming in iPS cells

Our ability to use induced pluripotent stem cells for cellular therapy requires that these cells have unrestricted proliferative capacity and that they maintain genome integrity during extensive periods in culture. Both of these requirements are directly associated with efficient telomerase activity and telomere homeostasis. In fact, consistent with a pluripotent state, initial reports of iPS cell generation showed that TERT was effectively upregulated after cellular reprogramming in both mouse[123] and human[122] cells. Moreover, reprogramming cells from a mouse expressing a GFP-TERT reporter, it was determined that the reactivation of TERT expression and telomerase activity was a late event during the cellular reprogramming process.[129] Moreover, the addition of TERT to the reprogramming "cocktail" (the four factors commonly used for iPS cell generation) increased reprogramming efficiency of human fibroblasts.[130] Confirming the importance of telomerase activity for cell reprogramming, it was shown that telomere elongation following cellular reprogramming is dependent on telomerase and not on alternative mechanisms of lengthening.[131] Likewise, the same authors show that although iPS cells were effectively generated from telomerase knockout mouse cells with long telomeres, the efficiency of reprogramming of telomerase-ablated cells with short telomeres is significantly reduced.[131] The low efficiency of cellular reprogramming in cells with short telomeres and other types of genetic damage can be rescued by abrogation of the p53 DNA damage response pathway, indicating that it could possibly work as a mechanism to ensure the genetic integrity of reprogrammed cells.[132–136] Giving further support for the importance of telomere homeostasis during cellular reprogramming, it has been shown that iPS cells with longer telomeres generate mouse chimeras with higher efficiency than those with short telomeres, when injected into blastocysts.[36] Interestingly, a recent paper compared the efficiency in cellular reprogramming of telomerase-ablated mouse cells, using either SCNT or induced pluripotency. Consistent with previous results described here earlier,[76] telomeres were elongated dramatically in cells reprogrammed using SCNT, due to a telomerase-independent mechanism. However, cells reprogrammed to an iPS state did not show telomere elongation during the process. This led to a higher differentiation potential and self-renewal capacity of cells generated by SCNT when compared to iPS cells.[137] Curiously, it has recently been shown that overexpression of catalytically active TERT is able to prevent X-chromosome skewing during cellular reprogramming, although the precise mechanism of TERT action in this case is still to be determined.[137] Taken together, the results presented here indicate the importance of proper telomere elongation during derivation of iPS cells and establish telomerase reactivation as a major component of cellular reprogramming.

3.2.2 Derivation of IPS cells harboring disease-associated telomerase mutations

The upregulation of TERT in wild-type mouse and human iPS cells leads to a progressive telomere elongation, resulting in telomeres that are similar in length to telomeres in ES cells.[131,138] In addition, as previously discussed, iPS cells derived from telomerase knockout mouse embryonic fibroblasts show little to none telomere elongation, indicating that telomerase activity is the driving force of telomere elongation and maintenance in iPS cells.[131,137] In human samples, the importance of telomerase activity for telomere maintenance and consequent self-renewal capability of iPS cells can be directly inferred from experiments using cells derived from patients harboring mutations in different components of telomerase and telomere-binding genes, such as DC and aplastic anemia. Since in these diseases, which are discussed in detail elsewhere in this book, patients have varying degrees of telomere attrition and stem cell impairment, the generation of patient-specific iPS cells would represent a unique model to understand the molecular physiology behind telomere shortening induced tissue failure.[139] Accordingly, different research groups tried to pursue this aim and have reported the successful generation of iPS cells derived from DC[140,141] and aplastic anemia patients.[142] Although cellular reprogramming from adult dermal fibroblasts derived from DC patients has proven to be extremely challenging, an optimized protocol where cells are kept and reprogrammed under low-oxygen conditions significantly increased reprogramming efficiency in those samples.[140] Using this improved protocol, the generation of iPS cells from patients carrying different mutations in DKC1, TCAB1, and TERT showed that telomere homeostasis is impaired in telomerase-mutant iPS cells. Surprisingly, it was shown that the stem cell-specific transcription factors OCT4 and NANOG bind to both TERC and dyskerin regulatory elements, thereby increasing their expression,[141] a previously unknown aspect of telomerase regulation in stem cells. Importantly, experiments with these cells also showed the importance of cellular reprogramming for accurate disease modeling *in vitro*. While TERT-mutant iPS cells showed marginal telomere dysfunction, TCAB1- and DKC1-mutant iPS cells presented with a severe telomere dysfunction phenotype, showing progressive telomere shortening that ultimately resulted in the loss of self-renewal of these cells. These results suggest that DC iPS cells might be mimicking the molecular events behind loss of self-renewal in adult stem cell compartments in DC patients, providing an explanation for the recently described correlation between disease severity and telomere length in patients afflicted with this disease.[143] Similar results were obtained from iPS cells derived from aplastic anemia patients, carrying mutations in TERT and TERC, which elongated telomeres at a lower rate when compared with wild-type iPS cells.[142] Importantly, telomerase-mutant iPS cells showed defective hematopoietic differentiation *in vitro*, therefore mimicking the clinical aspects of this disease in patients and demonstrating that human telomere diseases can be accurately modeled utilizing iPS cells.[142] When taken together, these results provide a clear genetic evidence for the importance of telomerase regulation and function in human pluripotent cells.[139]

4. CONCLUSIONS

Telomerase activity is tightly regulated in both mouse and human samples, and the limiting factor for this regulation seems to be TERT expression, which is limited to stem and progenitor cell populations. The importance of telomerase activity for stem cell function is clear in embryonic, adult, and induced pluripotent stem cells, where a failure to properly maintain telomere length leads to loss of self-renewal, the major feature of these cellular populations. The deleterious consequences of telomere-induced loss of self-renewal can be observed in organismal fitness, where it results in reduced tissue function. Additionally, we have discussed the role of TERT in different developmental pathways, independently from its role in telomere elongation. The molecular and biochemical understanding of these process is still in its infancy and is prone to contribute significantly for our knowledge of human health.

REFERENCES

1. Jaenisch R, Young R. Stem cells, the molecular circuitry of pluripotency and nuclear reprogramming. *Cell*. 2008;132(4):567–582.
2. Evans MJ, Kaufman MH. Establishment in culture of pluripotential cells from mouse embryos. *Nature*. 1981;292(5819):154–156.

3. Martin GR. Isolation of a pluripotent cell line from early mouse embryos cultured in medium conditioned by teratocarcinoma stem cells. *Proc Natl Acad Sci USA*. 1981;78(12):7634–7638.
4. Thomson JA, Itskovitz-Eldor J, Shapiro SS, et al. Embryonic stem cell lines derived from human blastocysts. *Science*. 1998;282(5391):1145–1147.
5. Chen L, Daley GQ. Molecular basis of pluripotency. *Hum Mol Genet*. 2008;17(R1): R23–R27.
6. Young RA. Control of the embryonic stem cell state. *Cell*. 2011;144(6):940–954.
7. Rossant J. Stem cells and early lineage development. *Cell*. 2008;132(4):527–531.
8. Silva J, Smith A. Capturing pluripotency. *Cell*. 2008;132(4):532–536.
9. Snippert HJ, Clevers H. Tracking adult stem cells. *EMBO Rep*. 2011;12(2):113–122.
10. Diehn M, Cho RW, Clarke MF. Therapeutic implications of the cancer stem cell hypothesis. *Semin Radiat Oncol*. 2009;19(2):78–86.
11. Sahin E, Depinho RA. Linking functional decline of telomeres, mitochondria and stem cells during ageing. *Nature*. 2010;464(7288):520–528.
12. Griffith JD, Comeau L, Rosenfield S, et al. Mammalian telomeres end in a large duplex loop. *Cell*. 1999;97(4):503–514.
13. Karlseder J, Broccoli D, Dai Y, Hardy S, de Lange T. p53- and ATM-dependent apoptosis induced by telomeres lacking TRF2. *Science*. 1999;283(5406):1321–1325.
14. Denchi EL, de Lange T. Protection of telomeres through independent control of ATM and ATR by TRF2 and POT1. *Nature*. 2007;448(7157):1068–1071.
15. Guo X, Deng Y, Lin Y, et al. Dysfunctional telomeres activate an ATM-ATR-dependent DNA damage response to suppress tumorigenesis. *EMBO J*. 2007;26(22):4709–4719.
16. Hockemeyer D, Palm W, Else T, et al. Telomere protection by mammalian Pot1 requires interaction with Tpp1. *Nat Struct Mol Biol*. 2007;14(8):754–761.
17. Sfeir A, de Lange T. Removal of shelterin reveals the telomere end-protection problem. *Science*. 2012;336(6081):593–597.
18. Giraud-Panis MJ, Teixeira MT, Geli V, Gilson E. CST meets shelterin to keep telomeres in check. *Mol Cell*. 2010;39(5):665–676.
19. Pinto AR, Li H, Nicholls C, Liu JP. Telomere protein complexes and interactions with telomerase in telomere maintenance. *Front Biosci (Landmark Ed)*. 2011;16:187–207.
20. Olovnikov AM. A theory of marginotomy. The incomplete copying of template margin in enzymic synthesis of polynucleotides and biological significance of the phenomenon. *J Theor Biol*. 1973;41(1):181–190.
21. Harley CB, Futcher AB, Greider CW. Telomeres shorten during ageing of human fibroblasts. *Nature*. 1990;345(6274):458–460.
22. Ballabeni A, Park IH, Zhao R, et al. Cell cycle adaptations of embryonic stem cells. *Proc Natl Acad Sci USA*. 2011;108(48):19252–19257.
23. Fujii-Yamamoto H, Kim JM, Arai K, Masai H. Cell cycle and developmental regulations of replication factors in mouse embryonic stem cells. *J Biol Chem*. 2005;280(13):12976–12987.
24. Ruiz S, Panopoulos AD, Herrerias A, et al. A high proliferation rate is required for cell reprogramming and maintenance of human embryonic stem cell identity. *Curr Biol*. 2011;21(1):45–52.
25. Conklin JF, Baker J, Sage J. The RB family is required for the self-renewal and survival of human embryonic stem cells. *Nat Commun*. 2012;3:1244.
26. Maimets T, Neganova I, Armstrong L, Lako M. Activation of p53 by nutlin leads to rapid differentiation of human embryonic stem cells. *Oncogene*. 2008;27(40):5277–5287.
27. Greider CW, Blackburn EH. The telomere terminal transferase of Tetrahymena is a ribonucleoprotein enzyme with two kinds of primer specificity. *Cell*. 1987;51(6):887–898.

28. Greider CW, Blackburn EH. Identification of a specific telomere terminal transferase activity in Tetrahymena extracts. *Cell*. 1985;43(2 Pt 1):405–413.
29. Morin GB. The human telomere terminal transferase enzyme is a ribonucleoprotein that synthesizes TTAGGG repeats. *Cell*. 1989;59(3):521–529.
30. Niida H, Matsumoto T, Satoh H, et al. Severe growth defect in mouse cells lacking the telomerase RNA component. *Nat Genet*. 1998;19(2):203–206.
31. Wang Y, Erdmann N, Giannone RJ, et al. An increase in telomere sister chromatid exchange in murine embryonic stem cells possessing critically shortened telomeres. *Proc Natl Acad Sci USA*. 2005;102(29):10256–10260.
32. Bailey SM, Brenneman MA, Goodwin EH. Frequent recombination in telomeric DNA may extend the proliferative life of telomerase-negative cells. *Nucleic Acids Res*. 2004;32(12):3743–3751.
33. Niida H, Shinkai Y, Hande MP, et al. Telomere maintenance in telomerase-deficient mouse embryonic stem cells: characterization of an amplified telomeric DNA. *Mol Cell Biol*. 2000;20(11):4115–4127.
34. Zalzman M, Falco G, Sharova LV, et al. Zscan4 regulates telomere elongation and genomic stability in ES cells. *Nature*. 2010;464(7290):858–863.
35. Amano T, Hirata T, Falco G, et al. Zscan4 restores the developmental potency of embryonic stem cells. *Nat Commun*. 2013;4:1966.
36. Huang J, Wang F, Okuka M, et al. Association of telomere length with authentic pluripotency of ES/iPS cells. *Cell Res*. 2011;21(5):779–792.
37. Armstrong L, Saretzki G, Peters H, et al. Overexpression of telomerase confers growth advantage, stress resistance, and enhanced differentiation of ESCs toward the hematopoietic lineage. *Stem Cells*. 2005;23(4):516–529.
38. Yang C, Przyborski S, Cooke MJ, et al. A key role for telomerase reverse transcriptase unit in modulating human embryonic stem cell proliferation, cell cycle dynamics, and in vitro differentiation. *Stem Cells*. 2008;26(4):850–863.
39. Venteicher AS, Abreu EB, Meng Z, et al. A human telomerase holoenzyme protein required for Cajal body localization and telomere synthesis. *Science*. 2009;323(5914):644–648.
40. Blackburn EH, Collins K. Telomerase: an RNP enzyme synthesizes DNA. *Cold Spring Harb Perspect Biol*. 2011;3(5).
41. Mitchell JR, Wood E, Collins K. A telomerase component is defective in the human disease dyskeratosis congenita. *Nature*. 1999;402(6761):551–555.
42. Wright WE, Piatyszek MA, Rainey WE, Byrd W, Shay JW. Telomerase activity in human germline and embryonic tissues and cells. *Dev Genet*. 1996;18(2):173–179.
43. Varela E, Schneider RP, Ortega S, Blasco MA. Different telomere-length dynamics at the inner cell mass versus established embryonic stem (ES) cells. *Proc Natl Acad Sci USA*. 2011;108(37):15207–15212.
44. Zeng S, Liu L, Sun Y, et al. Telomerase-mediated telomere elongation from human blastocysts to embryonic stem cells. *J Cell Sci*. 2013;127(Pt 4):752–762.
45. Saretzki G, Walter T, Atkinson S, et al. Downregulation of multiple stress defense mechanisms during differentiation of human embryonic stem cells. *Stem Cells*. 2008;26(2):455–464.
46. Yehezkel S, Rebibo-Sabbah A, Segev Y, et al. Reprogramming of telomeric regions during the generation of human induced pluripotent stem cells and subsequent differentiation into fibroblast-like derivatives. *Epigenetics*. 2011;6(1):63–75.
47. Fuchs E, Chen T. A matter of life and death: self-renewal in stem cells. *EMBO Rep*. 2013;14(1):39–48.
48. Li L, Clevers H. Coexistence of quiescent and active adult stem cells in mammals. *Science*. 2010;327(5965):542–545.
49. Buczacki SJ, Zecchini HI, Nicholson AM, et al. Intestinal label-retaining cells are secretory precursors expressing Lgr5. *Nature*. 2013;495(7439):65–69.

50. Clevers H. Stem cells: a unifying theory for the crypt. *Nature*. 2013;495(7439):53–54.
51. Morrison SJ, Prowse KR, Ho P, Weissman IL. Telomerase activity in hematopoietic cells is associated with self-renewal potential. *Immunity*. 1996;5(3):207–216.
52. Yui J, Chiu CP, Lansdorp PM. Telomerase activity in candidate stem cells from fetal liver and adult bone marrow. *Blood*. 1998;91(9):3255–3262.
53. Allsopp RC, Cheshier S, Weissman IL. Telomere shortening accompanies increased cell cycle activity during serial transplantation of hematopoietic stem cells. *J Exp Med*. 2001;193(8):917–924.
54. Vaziri H, Dragowska W, Allsopp RC, et al. Evidence for a mitotic clock in human hematopoietic stem cells: loss of telomeric DNA with age. *Proc Natl Acad Sci USA*. 1994;91(21):9857–9860.
55. Allsopp RC, Morin GB, DePinho R, Harley CB, Weissman IL. Telomerase is required to slow telomere shortening and extend replicative lifespan of HSCs during serial transplantation. *Blood*. 2003;102(2):517–520.
56. Allsopp RC, Morin GB, Horner JW, et al. Effect of TERT over-expression on the long-term transplantation capacity of hematopoietic stem cells. *Nat Med*. 2003;9(4):369–371.
57. Ramirez RD, Wright WE, Shay JW, Taylor RS. Telomerase activity concentrates in the mitotically active segments of human hair follicles. *J Invest Dermatol*. 1997;108(1):113–117.
58. Ueda M, Ouhtit A, Bito T, et al. Evidence for UV-associated activation of telomerase in human skin. *Cancer Res*. 1997;57(3):370–374.
59. Flores I, Cayuela ML, Blasco MA. Effects of telomerase and telomere length on epidermal stem cell behavior. *Science*. 2005;309(5738):1253–1256.
60. Sarin KY, Cheung P, Gilison D, et al. Conditional telomerase induction causes proliferation of hair follicle stem cells. *Nature*. 2005;436(7053):1048–1052.
61. Stange DE, Clevers H. Concise review: the yin and yang of intestinal (cancer) stem cells and their progenitors. *Stem Cells*. 2013;31(11):2287–2295.
62. Leblond CP, Stevens CE. The constant renewal of the intestinal epithelium in the albino rat. *Anat Rec*. 1948;100(3):357–377.
63. Breault DT, Min IM, Carlone DL, et al. Generation of mTert-GFP mice as a model to identify and study tissue progenitor cells. *Proc Natl Acad Sci USA*. 2008;105(30):10420–10425.
64. Montgomery RK, Carlone DL, Richmond CA, et al. Mouse telomerase reverse transcriptase (mTert) expression marks slowly cycling intestinal stem cells. *Proc Natl Acad Sci USA*. 2011;108(1):179–184.
65. Schepers AG, Vries R, van den Born M, van de Wetering M, Clevers H. Lgr5 intestinal stem cells have high telomerase activity and randomly segregate their chromosomes. *EMBO J*. 2011;30(6):1104–1109.
66. Pech MF, Artandi SE. TRAPping telomerase within the intestinal stem cell niche. *EMBO J*. 2011;30(6):986–987.
67. Sacco A, Mourkioti F, Tran R, et al. Short telomeres and stem cell exhaustion model Duchenne muscular dystrophy in mdx/mTR mice. *Cell*. 2010;143(7):1059–1071.
68. Mourkioti F, Kustan J, Kraft P, et al. Role of telomere dysfunction in cardiac failure in Duchenne muscular dystrophy. *Nat Cell Biol*. 2013;15(8):895–904.
69. Gage FH, Temple S. Neural stem cells: generating and regenerating the brain. *Neuron*. 2013;80(3):588–601.
70. Wang S, Chandler-Militello D, Lu G, et al. Prospective identification, isolation, and profiling of a telomerase-expressing subpopulation of human neural stem cells, using sox2 enhancer-directed fluorescence-activated cell sorting. *J Neurosci*. 2010;30(44):14635–14648.

71. Ferron S, Mira H, Franco S, et al. Telomere shortening and chromosomal instability abrogates proliferation of adult but not embryonic neural stem cells. *Development*. 2004;131(16):4059–4070.
72. Eisenberg DT, Hayes MG, Kuzawa CW. Delayed paternal age of reproduction in humans is associated with longer telomeres across two generations of descendants. *Proc Natl Acad Sci USA*. 2012;109(26):10251–10256.
73. Maki CB, Pacchiarotti J, Ramos T, et al. Phenotypic and molecular characterization of spermatogonial stem cells in adult primate testes. *Hum Reprod*. 2009;24(6):1480–1491.
74. Izadyar F, Wong J, Maki C, et al. Identification and characterization of repopulating spermatogonial stem cells from the adult human testis. *Hum Reprod*. 2011;26(6):1296–1306.
75. Yan L, Wu S, Zhang S, Ji G, Gu A. Genetic variants in telomerase reverse transcriptase (TERT) and telomerase-associated protein 1 (TEP1) and the risk of male infertility. *Gene*. 2014;534(2):139–143.
76. Liu L, Bailey SM, Okuka M, et al. Telomere lengthening early in development. *Nat Cell Biol*. 2007;9(12):1436–1441.
77. Schaetzlein S, Lucas-Hahn A, Lemme E, et al. Telomere length is reset during early mammalian embryogenesis. *Proc Natl Acad Sci USA*. 2004;101(21):8034–8038.
78. Blasco MA, Lee HW, Hande MP, et al. Telomere shortening and tumor formation by mouse cells lacking telomerase RNA. *Cell*. 1997;91(1):25–34.
79. Rudolph KL, Chang S, Lee HW, et al. Longevity, stress response, and cancer in aging telomerase-deficient mice. *Cell*. 1999;96(5):701–712.
80. Lee HW, Blasco MA, Gottlieb GJ, et al. Essential role of mouse telomerase in highly proliferative organs. *Nature*. 1998;392(6676):569–574.
81. Beamer WG, Sweet HO, Bronson RT, et al. Adrenocortical dysplasia: a mouse model system for adrenocortical insufficiency. *J Endocrinol*. 1994;141(1):33–43.
82. Keegan CE, Hutz JE, Else T, et al. Urogenital and caudal dysgenesis in adrenocortical dysplasia (acd) mice is caused by a splicing mutation in a novel telomeric regulator. *Hum Mol Genet*. 2005;14(1):113–123.
83. Hockemeyer D, Palm W, Wang RC, Couto SS, de Lange T. Engineered telomere degradation models dyskeratosis congenita. *Genes Dev*. 2008;22(13):1773–1785.
84. Wang Y, Shen MF, Chang S. Essential roles for Pot1b in HSC self-renewal and survival. *Blood*. 2011;118(23):6068–6077.
85. Savage SA, Alter BP. Dyskeratosis congenita. *Hematol Oncol Clin North Am*. 2009;23(2):215–231.
86. Bessler M, Wilson DB, Mason PJ. Dyskeratosis congenita. *FEBS Lett*. 2010;584(17):3831–3838.
87. Armanios M, Blackburn EH. The telomere syndromes. *Nat Rev Genet*. 2012;13(10):693–704.
88. Armanios M. Syndromes of telomere shortening. *Annu Rev Genomics Hum Genet*. 2009;10:45–61.
89. Imamura S, Uchiyama J, Koshimizu E, et al. A non-canonical function of zebrafish telomerase reverse transcriptase is required for developmental hematopoiesis. *PLoS One*. 2008;3(10):e3364.
90. Choi J, Southworth LK, Sarin KY, et al. TERT promotes epithelial proliferation through transcriptional control of a Myc- and Wnt-related developmental program. *PLoS Genet*. 2008;4(1):e10.
91. Park JI, Venteicher AS, Hong JY, et al. Telomerase modulates Wnt signalling by association with target gene chromatin. *Nature*. 2009;460(7251):66–72.
92. Hoffmeyer K, Raggioli A, Rudloff S, et al. Wnt/beta-catenin signaling regulates telomerase in stem cells and cancer cells. *Science*. 2012;336(6088):1549–1554.

93. Strong MA, Vidal-Cardenas SL, Karim B, et al. Phenotypes in mTERT(+)/(−) and mTERT(−)/(−) mice are due to short telomeres, not telomere-independent functions of telomerase reverse transcriptase. *Mol Cell Biol.* 2011;31(12):2369–2379.
94. Listerman I, Gazzaniga FS, Blackburn EH. An investigation of the effects of the core protein telomerase reverse transcriptase on Wnt signaling in breast cancer cells. *Mol Cell Biol.* 2014;34(2):280–289.
95. Shkreli M, Sarin KY, Pech MF, et al. Reversible cell-cycle entry in adult kidney podocytes through regulated control of telomerase and Wnt signaling. *Nat Med.* 2012;18(1):111–119.
96. Maida Y, Yasukawa M, Furuuchi M, et al. An RNA-dependent RNA polymerase formed by TERT and the RMRP RNA. *Nature.* 2009;461(7261):230–235.
97. Ridanpaa M, van Eenennaam H, Pelin K, et al. Mutations in the RNA component of RNase MRP cause a pleiotropic human disease, cartilage–hair hypoplasia. *Cell.* 2001;104(2):195–203.
98. Hermanns P, Bertuch AA, Bertin TK, et al. Consequences of mutations in the non-coding RMRP RNA in cartilage–hair hypoplasia. *Hum Mol Genet.* 2005;14(23):3723–3740.
99. Rosenbluh J, Nijhawan D, Chen Z, et al. RMRP is a non-coding RNA essential for early murine development. *PLoS One.* 2011;6(10):e26270.
100. Briggs R, King TJ. Transplantation of living nuclei from blastula cells into enucleated frogs' eggs. *Proc Natl Acad Sci USA.* 1952;38(5):455–463.
101. Gurdon JB. The developmental capacity of nuclei taken from intestinal epithelium cells of feeding tadpoles. *J Embryol Exp Morphol.* 1962;10:622–640.
102. Gurdon JB, Laskey RA, Reeves OR. The developmental capacity of nuclei transplanted from keratinized skin cells of adult frogs. *J Embryol Exp Morphol.* 1975;34(1):93–112.
103. Wilmut I, Schnieke AE, McWhir J, Kind AJ, Campbell KH. Viable offspring derived from fetal and adult mammalian cells. *Nature.* 1997;385(6619):810–813.
104. Jaenisch R. Stem cells, pluripotency and nuclear reprogramming. *J Thromb Haemost.* 2009;7(Suppl 1):21–23.
105. Eggan K, Baldwin K, Tackett M, et al. Mice cloned from olfactory sensory neurons. *Nature.* 2004;428(6978):44–49.
106. Hochedlinger K, Jaenisch R. Monoclonal mice generated by nuclear transfer from mature B and T donor cells. *Nature.* 2002;415(6875):1035–1038.
107. Blau HM, Baltimore D. Differentiation requires continuous regulation. *J Cell Biol.* 1991;112(5):781–783.
108. Blau HM, Chiu CP, Webster C. Cytoplasmic activation of human nuclear genes in stable heterocaryons. *Cell.* 1983;32(4):1171–1180.
109. Blau HM, Pavlath GK, Hardeman EC, et al. Plasticity of the differentiated state. *Science.* 1985;230(4727):758–766.
110. Johansson CB, Youssef S, Koleckar K, et al. Extensive fusion of haematopoietic cells with Purkinje neurons in response to chronic inflammation. *Nat Cell Biol.* 2008;10(5):575–583.
111. Weimann JM, Charlton CA, Brazelton TR, Hackman RC, Blau HM. Contribution of transplanted bone marrow cells to Purkinje neurons in human adult brains. *Proc Natl Acad Sci USA.* 2003;100(4):2088–2093.
112. Yamanaka S, Blau HM. Nuclear reprogramming to a pluripotent state by three approaches. *Nature.* 2010;465(7299):704–712.
113. Shiels PG, Kind AJ, Campbell KH, et al. Analysis of telomere lengths in cloned sheep. *Nature.* 1999;399(6734):316–317.
114. Wakayama T, Shinkai Y, Tamashiro KL, et al. Cloning of mice to six generations. *Nature.* 2000;407(6802):318–319.

115. Lanza RP, Cibelli JB, Blackwell C, et al. Extension of cell life-span and telomere length in animals cloned from senescent somatic cells. *Science*. 2000;288(5466):665–669.
116. Tian XC, Xu J, Yang X. Normal telomere lengths found in cloned cattle. *Nat Genet*. 2000;26(3):272–273.
117. Betts D, Bordignon V, Hill J, et al. Reprogramming of telomerase activity and rebuilding of telomere length in cloned cattle. *Proc Natl Acad Sci USA*. 2001;98(3):1077–1082.
118. Miyashita N, Kubo Y, Yonai M, et al. Cloned cows with short telomeres deliver healthy offspring with normal-length telomeres. *J Reprod Dev*. 2011;57(5):636–642.
119. Clark AJ, Ferrier P, Aslam S, et al. Proliferative lifespan is conserved after nuclear transfer. *Nat Cell Biol*. 2003;5(6):535–538.
120. Xu J, Yang X. Telomerase activity in early bovine embryos derived from parthenogenetic activation and nuclear transfer. *Biol Reprod*. 2001;64(3):770–774.
121. Dang-Nguyen TQ, Haraguchi S, Akagi S, et al. Telomere elongation during morula-to-blastocyst transition in cloned porcine embryos. *Cell Reprogram*. 2012;14(6):514–519.
122. Takahashi K, Tanabe K, Ohnuki M, et al. Induction of pluripotent stem cells from adult human fibroblasts by defined factors. *Cell*. 2007;131(5):861–872.
123. Takahashi K, Yamanaka S. Induction of pluripotent stem cells from mouse embryonic and adult fibroblast cultures by defined factors. *Cell*. 2006;126(4):663–676.
124. Yu J, Vodyanik MA, Smuga-Otto K, et al. Induced pluripotent stem cell lines derived from human somatic cells. *Science*. 2007;318(5858):1917–1920.
125. Warren L, Manos PD, Ahfeldt T, et al. Highly efficient reprogramming to pluripotency and directed differentiation of human cells with synthetic modified mRNA. *Cell Stem Cell*. 2010;7(5):618–630.
126. Anokye-Danso F, Trivedi CM, Juhr D, et al. Highly efficient miRNA-mediated reprogramming of mouse and human somatic cells to pluripotency. *Cell Stem Cell*. 2011;8(4):376–388.
127. Zhou H, Wu S, Joo JY, et al. Generation of induced pluripotent stem cells using recombinant proteins. *Cell Stem Cell*. 2009;4(5):381–384.
128. Robinton DA, Daley GQ. The promise of induced pluripotent stem cells in research and therapy. *Nature*. 2012;481(7381):295–305.
129. Stadtfeld M, Maherali N, Breault DT, Hochedlinger K. Defining molecular cornerstones during fibroblast to iPS cell reprogramming in mouse. *Cell Stem Cell*. 2008;2(3):230–240.
130. Park IH, Zhao R, West JA, et al. Reprogramming of human somatic cells to pluripotency with defined factors. *Nature*. 2008;451(7175):141–146.
131. Marion RM, Strati K, Li H, et al. Telomeres acquire embryonic stem cell characteristics in induced pluripotent stem cells. *Cell Stem Cell*. 2009;4(2):141–154.
132. Utikal J, Polo JM, Stadtfeld M, et al. Immortalization eliminates a roadblock during cellular reprogramming into iPS cells. *Nature*. 2009;460(7259):1145–1148.
133. Marion RM, Strati K, Li H, et al. A p53-mediated DNA damage response limits reprogramming to ensure iPS cell genomic integrity. *Nature*. 2009;460(7259):1149–1153.
134. Li H, Collado M, Villasante A, et al. The Ink4/Arf locus is a barrier for iPS cell reprogramming. *Nature*. 2009;460(7259):1136–1139.
135. Kawamura T, Suzuki J, Wang YV, et al. Linking the p53 tumour suppressor pathway to somatic cell reprogramming. *Nature*. 2009;460(7259):1140–1144.
136. Hong H, Takahashi K, Ichisaka T, et al. Suppression of induced pluripotent stem cell generation by the p53-p21 pathway. *Nature*. 2009;460(7259):1132–1135.
137. Le R, Kou Z, Jiang Y, et al. Enhanced telomere rejuvenation in pluripotent cells reprogrammed via nuclear transfer relative to induced pluripotent stem cells. *Cell Stem Cell*. 2014;14(1):27–39.

138. Suhr ST, Chang EA, Rodriguez RM, et al. Telomere dynamics in human cells reprogrammed to pluripotency. *PLoS One*. 2009;4(12):e8124.
139. Batista LF, Artandi SE. Understanding telomere diseases through analysis of patient-derived iPS cells. *Curr Opin Genet Dev*. 2013;23(5):526–533.
140. Batista LF, Pech MF, Zhong FL, et al. Telomere shortening and loss of self-renewal in dyskeratosis congenita induced pluripotent stem cells. *Nature*. 2011;474(7351):399–402.
141. Agarwal S, Loh YH, McLoughlin EM, et al. Telomere elongation in induced pluripotent stem cells from dyskeratosis congenita patients. *Nature*. 2010;464(7286):292–296.
142. Winkler T, Hong SG, Decker JE, et al. Defective telomere elongation and hematopoiesis from telomerase-mutant aplastic anemia iPSCs. *J Clin Invest*. 2013;123(5):1952–1963.
143. Alter BP, Rosenberg PS, Giri N, et al. Telomere length is associated with disease severity and declines with age in dyskeratosis congenita. *Haematologica*. 2012;97(3):353–359.

CHAPTER FOUR

Telomere Dynamics and Aging

Geraldine Aubert
Terry Fox Laboratory, British Columbia Cancer Agency, Vancouver, British Columbia, Canada

Contents

Abstract

Telomeres consist of repetitive DNA–protein complexes that cap the ends of vertebrate linear chromosomes. Their capping function and dynamics both with regard to structure and length are carefully orchestrated by many regulatory mechanisms and factors, with likely more yet to be described. Telomere shortening has been shown to be a major measurable molecular characteristic of aging of cells *in vitro* and *in vivo* and is thought to have evolved as a tumor protection mechanism in long-lived species. Regulators and modifiers of telomere dynamics and dynamics with age together with the consequences of telomere shortening and telomere dysfunction in the context of aging and aging-related disorders are discussed in this chapter.

Progress in Molecular Biology and Translational Science, Volume 125
ISSN 1877-1173
http://dx.doi.org/10.1016/B978-0-12-397898-1.00004-9

1. INTRODUCTION: TELOMERE STRUCTURE, DETERMINANTS OF TELOMERE DYNAMICS, MEASUREMENT PARAMETERS, AND REGULATION MECHANISMS

1.1. About this chapter

Telomeres are the natural ends of linear chromosomes and are composed of $(TTAGGG)_n$ repeat tracks in vertebrates including humans.[1,2] They form protective structures that differentiate telomeric DNA from exposed or damaged DNA and prevent their recognition as breaks by the DNA repair machinery. The importance of telomeres to health and disease and that of the telomerase holoenzyme have been highlighted since their discovery.[3] Telomere and telomerase dynamics have been connected to the proliferative potential of the cell that has in turn been likened to cellular aging since it is limited. These have also been implicated in the changes and disease mechanisms that are part of the aging process on the molecular and cellular levels, as well as at the level of the whole organism. Since aging and aging-associated pathologies are complex multifactorial phenomena that occur in continuity and evolve over time, this chapter will focus on aspects of aging throughout the human life span for which the evidence of the involvement of telomere biology has been demonstrated. How aging affects telomere dynamics and associating factors has been the focus of much research related to human health, which inherently declines with increased age. However, some determining parameters may be set early on in life, which will also be discussed. This fairly wide topic is the subject of this chapter, which is closely connected to all other chapters in this publication. Some of the salient points mentioned for clarity in this chapter will be developed in further details by my colleagues in this publication, and the reader will be referred to these sections accordingly. As the body of works in this area of biology and medicine is becoming very consequent, many excellent works will unfortunately not be covered in this chapter due to space constraints.

1.2. Telomere structure and determinants of telomere dynamics

Telomere biology has been intimately connected and highlighted as one of the major pillars or hallmarks of cancer[4,5] and of aging,[6] respectively. Both are interconnected through the development of genomic instability and are mediated in part by telomere function and homeostasis. The main function of telomeres is to act as a protective system so as to ensure the integrity of individual chromosomes and their accurate segregation during events including meiosis and mitosis. The repetitive nature and sequence of telomere DNA are targeted by specific proteins that bind the DNA directly, as well as an interacting scaffold of proteins. This six-protein complex is collectively termed "shelterin"; for the shielding function, they serve to effectively hide the telomeric DNA from repair mechanisms that would recognize an unbound DNA end as an exposed sequence to be targeted.[7–9] These include the proteins telomere repeat-binding factors TRF1 and TRF2 that directly bind the double-stranded telomere repeats and protection of telomeres 1 (POT1) that binds to the 3′ single-stranded G-rich overhang present at the very end of chromosomes. Other telomere-specific proteins that interact with TRF1, TRF2, and POT1 include RAP1 (also described as TRF2-interacting protein), TIN2–POT1-interacting protein 1 TPP1 (encoded by the *ACD* gene), and TRF1-interacting nuclear protein 2 or TIN2, as well as the Ku70/80 complex present at the junction of single- and double-stranded telomeric DNA. TRF1 and TRF2 are known to bend telomeric DNA[10,11] and, if the length of telomere repeat track is sufficiently long, allow for 3′ single-stranded invasion into duplex telomere DNA and mediate the formation of the telomere loop or T-loop[12] (Fig. 4.1A).

Further details about telomere repeat-interacting proteins and their basic function and interactions are described in Chapter 1. The presence of the T-loop has been proposed to block the progress of the DNA replication machinery and prevent telomerase accessibility and loading onto the 3′ G-rich overhang in order to add *de novo* telomere repeats. The resolution of the T-loop structure has therefore been postulated to be required for DNA replication. The process of DNA replication itself cannot include the entire length of the telomere track on the lagging strand due to the requirement of RNA primers for lagging strand replication. The last-positioned primer along the telomere track is removed to ligate an Okazaki fragment and form the newly synthesized strand, thereby leaving a primer length segment unreplicated. This is referred to as the end replication problem[13] and inherently results in telomere shortening with each cell division (Fig. 4.1B). Further, telomere ends are processed by exonucleases (ExoI and Apollo) also leading to shortening of the 5′ end in order to reconstitute the overhang structure.[14] 5′ Fill-in is mediated by the CST complex[15,16] (CTC1, STN1, and TEN1), which can counteract the effect of exonucleases and can be considered one of the telomere maintenance mechanisms.

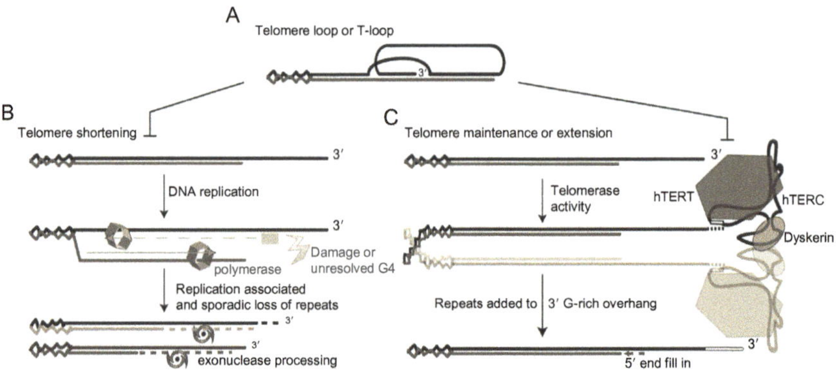

Figure 4.1 Telomere length shortening and maintenance mechanisms. (A) Schematic representation of a telomere loop or T-loop formed by strand invasion of the G-rich telomere strand overhang into the double-stranded telomere DNA repeats. The formation of this structure is mediated by shelterin telomere-binding proteins (not shown, see Chapter 1) and forms a protective cap at the chromosome end. Resolution of the T-loop structure is required for DNA replication and catalytic telomere repeat addition by telomerase. (B) Main telomere shortening mechanisms are inherent to cell division and include the end replication problem whereby the DNA replication machinery is not able to replicate to the very end of a linear DNA template on the lagging strand (box). Damage to DNA or residual higher-order structures during replication can lead to sporadic loss of telomere repeats (which can include the entire T-loop). Further, telomere ends are resected by Apollo and ExoI exonucleases in order to restore the G-rich overhang following replication, allowing for the protective binding of shelterin telomere-binding proteins and the formation of a T-loop under specific circumstances. This process is dependent upon telomere length and is controlled by shelterin proteins (see Chapter 1). (C) The sole (nonrecombination-mediated) telomere maintenance mechanism is through the action of the telomerase holoenzyme, composed of three minimal components hTERT, hTERC, and dyskerin, adding TTAGGG repeats to the 3′ G-rich overhang. Telomerase acts as a dimer and may allow simultaneous repeat addition on sister telomere ends (depicted as a dimmer mirror image). Telomerase is absent or below detectable levels in most somatic cells. When expressed, telomerase levels are limiting and tightly regulated making telomere shortening mechanisms dominant. Although 5′ fill-in occurs as part of telomere length maintenance at the end of DNA replication (by the CST complex), it does not result in a net gain in telomere repeats as measured by the majority of methods conventionally used.

Telomere-binding proteins have been shown to transiently come apart from telomeric DNA during the progression of S to G2 phase of the cell cycle[17] and resume their binding and protective functions thereafter. Such dynamics may also result in temporary exposure of DNA to potentially cell intrinsic damaging agents such as reactive oxygen species (ROS) and cell extrinsic damaging agents such as UV and radiation or may lead to the formation of higher-order DNA structures such as G4 DNA (Fig. 4.1B). These can impair or stall the progress of the replication machinery, can trigger DNA damage response pathways to deal with replication problems, and could lead to deletion or repair events leading to telomere shortening. Telomere-trimming events have also been shown to occur, potentially to control the upper limit of telomere length, and can result in the cleavage of the entire T-loop from the telomere end, generating a T-circle molecule.[18–20]

Since T-loop formation is dependent upon the length of telomere DNA track, access by end processing factors and telomerase has been proposed to be blocked unless the telomere conformation is linear. Telomerase is a reverse transcriptase holoenzyme[3] requiring three minimal essential components for catalytic activity: a reverse transcriptase protein (TERT),[21,22] a nontranscribed RNA template molecule (TERC also referred to as TR),[23] and the H/ACA snRNP member dyskerin (DKC1).[24,25] Other small nuclear RiboNucleoprotein Particles (snRNPs) are also known to participate in the telomerase complex, and though they are not essential to the *in vitro* catalytic activity of telomerase (see Chapter 1), they play an important role in telomere maintenance *in vivo* as demonstrated by their causative involvement in telomere deficiency syndromes (see Chapters 6, 7 and 8). Importantly, TERT is the only essential human telomerase component that is not ubiquitously expressed (by contrast, murine TERT is ubiquitously expressed). Telomerase acts as a dimer that may extend sister telomeres simultaneously[26,27] (shown as a mirror image, Fig. 4.1C). The telomerase holoenzyme is assembled and folded in Cajal bodies and shuttles from these defined nuclear locations to telomere ends via recruitment interactions with the proteins TCAB1[28,29] and HOT1[30] followed by loading and stimulation of telomerase activity and processivity through the shelterin proteins TPP1 and POT1.[31,32] Telomerase access to telomeres is regulated by both telomere-specific and transient protein and RNA factors. For example, the CST complex has recently been shown to mediate telomerase shutoff on the G-rich telomere overhang.[33] Furthermore, more evidence is needed to fully elucidate conflicting reports on the influence of the transcribed but nontranslated telomere track RNA molecules or telomeric repeat-containing RNA (TERRA) on telomere homeostasis and telomerase activity in different systems.[34–36] In addition, telomeres are subjected to similar chromatin compaction to the rest of the genome (reviewed in Ref. 37) and have specific functions in sister telomere cohesion.[38] Both these processes affect telomerase access to telomeric DNA.

Collectively, multiple networks orchestrate telomerase expression levels, posttranscriptional modifications (reviewed in Ref. 39), the accessibility of the telomere to the DNA replication machinery, the DNA repair machinery, and telomere repeat addition capacity by telomerase. The regulated expression levels, splicing, cellular trafficking, and mediators thereof for any of these molecules and networks influence telomere dynamics.

As new factors and contributing factors are being uncovered, the complexity and orchestration of telomere dynamics and telomere length dynamics are being delineated, and it is likely their contribution to health and disease states will warrant further research. A major challenge remains in translating how a change in any one of these factors impacts the aging process and aging-related pathologies, when they may be interdependent with additional factors.

1.3. Measurement parameters, regulation mechanisms, natural variability, and natural variants

The status of telomere integrity is related to both telomere length and telomerase catalytic activity. Therefore, these are the two major parameters that are evaluated in cell biology studies and in primary tissue samples that can be evaluated in the clinical setting. A detailed review of telomere length measurement methods available has been described elsewhere previously.[40] The most studied tissue thus far with respect to telomere length in relation to aging is blood. This is due to its relative ease of sampling as well as minimal invasive process and due to the high-quality genetic material that can be obtained and tested from nucleated blood cells. This allows for overall measurement of nucleated cells (from whole blood) and of specific blood cell subsets together with immunostaining or purification strategies or DNA extracted from whole blood or from specific cell subsets. The separation of cells of interest in order to obtain a targeted measurement likely improves accuracy. Natural variability or range of telomere length distribution has been shown to be high between individuals.[41] Blood or leukocyte telomere length measurements reflect the turnover and the remaining proliferative capacity of cells responsible for hematopoiesis: the hematopoietic stem cell and cells of the immune system. Other tissues besides blood that have been more extensively studied in humans include skin fibroblasts, colon cells, muscle cells, fat cells, and oral epithelial cells. Modulators of telomere dynamics were described in the paragraph above and are major determinants of telomere dynamics. The other major determinant of telomere dynamics of tissues with age is the rate of turnover (cell division or mitotic index) or regeneration (renewal) that the tissue undergoes. Despite tissue-specific differences, relative telomere length synchrony exists between different tissues of the same individual,[42,43] and inherited telomere maintenance deficiencies have also been shown to affect all tissues of the same individual in a synchronous manner.[44] Due to these observations, telomere length in blood leucocytes is considered a good general surrogate measure of telomere length in other tissues of the same individual provided no acute or chronic condition has affected select tissue compartments. Of note, pathologies may not affect all tissue compartments or all cell types within a tissue compartment equally. Further, a cell type-specific biology (cell intrinsic) and turnover rate (cell extrinsic) may affect its telomere length relative to that of other cell types within the same tissue or other tissues from the same individual. For an example of cell intrinsic differential biology, breast luminal cells have recently been shown to have relatively short telomere length likely mediated by inherently high ROS levels.[45] In an example of cell extrinsic telomere length shortening, mitotic stress or high proliferative stimulation rate in order to regenerate hematopoiesis after a bone marrow transplant procedure (or hematopoietic stem cell transplant) gives rise to temporary accelerated telomere shortening.[46]

Telomerase activity constitutes the sole major positive telomere length regulation mechanism. Telomerase access to the telomere end is restricted by the presence of the T-loop and very closely regulated at the protein expression level under steady state healthy conditions[47] (Fig. 4.1). Telomerase expression is detected in germline and stem cells and is below detection level or absent in the majority of somatic cells. As a consequence, net telomere shortening occurs with each somatic cell division and is estimated to be between 50 and 100 bp. Exceptions of somatic cells expressing telomerase do include cells of the immune system that require telomerase activity to maintain proliferative potential and support selective and rapid transient clonal expansion (followed by contraction). These cells are present in blood and can be stimulated *in vitro* to induce telomerase expression[47] and ensuing catalytic activity. Multiple levels of telomerase regulation have been demonstrated and include transcription regulation, alternative splicing (generating catalytically inactive splice variants), holoenzyme complex formation, cellular localization, and telomere loading. Telomerase activity can be tested from cell extracts by *in vitro* biochemical assays that quantify *de novo* added repeats to a synthetic telomere oligonucleotide target. Although immune cells are able to express significant levels of the telomerase holoenzyme, these levels are not sufficient to induce telomere elongation *in vitro* or *in vivo* but may contribute to the maintenance of telomere length and support continuing proliferation. Therefore, even these cell types are not entirely shielded from the proliferation-dependent telomere shortening that is characteristic of somatic cells. Somatic cell telomere lengthening can be achieved *in vitro* by ectopic overexpression of telomerase and has only been described *in vivo* in a single report in the setting of aplastic anemia treatment.[48] Natural genetic variants of telomerase single nucleotide polymorphisms (SNPs) may result in different telomerase expression levels or different catalytic capabilities between individuals,[49,50] and such variants are being evaluated as potential disease risk factors (see also Section 4 and Chapters 2 and 5).

2. TELOMERE CONNECTION TO THE CHARACTERISTICS OF AGING: CELL SENESCENCE AND APOPTOSIS, MITOCHONDRIAL FUNCTION, AND METABOLISM

2.1. Senescence or apoptosis telomere checkpoint

The first observation that cellular division was a limited process was made by Hayflick who noted that consecutive cell proliferation *in vitro* and *in vivo* was leading to a slowing down of replicative kinetics, followed by the complete inability to progress through further cell divisions in a process that was accompanied by morphological, biochemical, and molecular changes. This is referred to as the Hayflick limit[51] or replicative senescence, reviewed in Ref. 52. Telomere loss with cell division was demonstrated in fibroblasts[53,54] and was shown to progress with age in blood and the colon[55] as well as in other tissues (reviewed in Ref. 56). Progressive telomere loss is also referred to as the cellular mitotic clock or telomere-counting mechanism; such telomere repeat loss has been proposed to have evolved as a tumor suppressor mechanism in long-lived species, effectively stopping cells that could progress towards potentially oncogenic genomic instability from further divisions (see Section 4 and Chapters 2 and 5). Cross-sectional leukocyte telomere length data between birth and age 100 years of age are depicted in Fig. 4.2.

The cross-sectional leukocyte data suggest that telomere length fits a model of gradual linear decline with age in adults, without significant changes in dynamics.[41] Perhaps surprisingly (considering not all tissues have comparable replication and regeneration rates), telomere dynamics of different tissue types isolated from a same adult individual appear to decline at a similar rate with age regardless of the tissue of origin.[57] Assuming measurement variables are equal between tissues, this observation supports the concept that selective telomerase expression may compensate for additional required

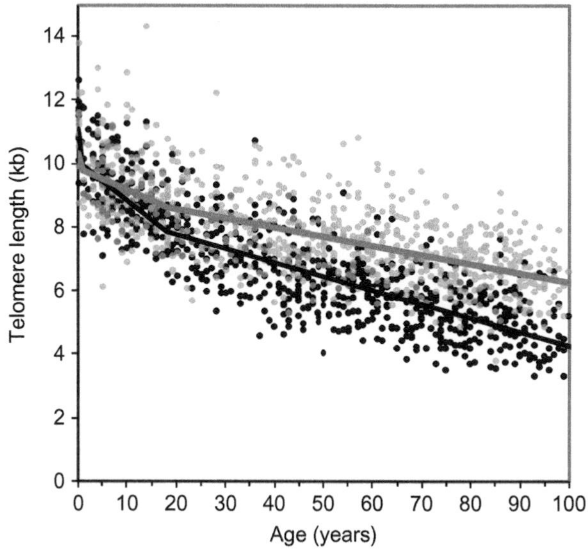

Figure 4.2 Leukocyte telomere dynamics' cross-sectional data with age. Cross-sectional data depicting leukocyte telomere length measurements according to age from 835 healthy individuals from birth (cord blood sample) and representing each age year with on average eight individuals per year to age 100 years. Two leukocyte cell subsets are depicted: granulocytes (lighter color) and lymphocytes (black), both present together in peripheral blood samples, to illustrate that the lymphocyte telomere length dynamics with age differ depending on the cell subset considered. The range of distribution is wide between individuals of a same age and appears maintained throughout life. *Adapted from Ref. 41.*

proliferative potential in high turnover tissues such as blood immune cells. It also suggests that the set point for a specific tissue entering adulthood could be a susceptibility factor over time for the development of tissue failure syndromes or genomic instability (in the lung, e.g., which like blood is also a high turnover tissue but where telomerase activity is not detected).

Replicative senescence is reached by progressive telomere loss that triggers a cell division checkpoint controlled by the ATM/p53 pathway that can lead to apoptosis or cell death.[58] Five dysfunctional telomeres in a single nucleus have been shown to be a critical threshold to senescence in human fibroblasts.[59] Senescence and apoptosis associated with short telomeres have been shown to also be associated with the loss of stem cells and the lack of regenerative capacity resulting in the loss of tissue function with aging. Tissue of origin is an important consideration, because the specific tissue turnover rate and ensuing telomere shortening will impact the integrity and function of the tissue. Since most blood-derived cells undergo apoptosis and are cleared rapidly, these would be eliminated from peripheral blood circulation and would no longer be accounted for. By contrast, for skin fibroblast that reaches senescence, cell clearance may not be as rapid and may lead to the accumulation of senescent cells within a tissue and create a proinflammatory environment.[60] Subsequent removal of senescent cells is mediated by immune cells and can be accompanied by scarring (fibrosis in some settings) and tissue remodeling or can lead to the support of tumor growth by initiating proinflammatory and tumor-promoting secreted factors (see Section 4 and Chapters 6–8). Telomere elongation in longitudinal aging studies has been described for limited periods of time on a human life span scale (up to 10 years) and has been attributed to experimental error.[61] More longitudinal data are needed to assess the impact of telomere regulators and to overcome challenges in data interpretation, which can be challenging due to a wide range between individuals of the same chronological age, and technical telomere length measurement accuracy parameters.[40]

Telomere shortening with age results in senescence or apoptosis in the majority of cells that reach the critical checkpoint leading to genomic instability (and increased cancer incidence), immunosenescence, and tissue failure syndrome phenotypes in a progressive process that fits the stem cell-deficit hypothesis model of aging. These causal connections are best illustrated by mouse model data demonstrating the progression between these phenotypes[62,63] and by the study of individuals with inherited telomerase haploinsufficiency that present with symptoms that recapitulate the phenotypes of stem cell deficit (albeit with variable genetic penetrance). The latter will be described in more detail in Section 4.

2.2. Telomeres, mitochondrial function, and metabolism

In addition to triggering critical cell cycle checkpoints, telomere dysfunction has been shown in mouse model systems to induce defective mitochondrial biogenesis and metabolic compromise through glucose-sensing problems.[64] Interestingly, a separate model using mitochondrial DNA mutator mice (which accumulate many point mutations due to a mitochondrial polymerase proofreading deficiency[65]; mitochondrial DNA mutations have been shown to accumulate with age) showed that these animals had fewer stem cells and had reduced self-renewal activity (reviewed in Ref. 66). Separate lines of evidence using bioinformatics comparative algorithm tools are pointing to similarities between normal human aging, mitochondrial diseases, and some accelerated aging disorders that display common clinical phenotypes.[67] Collectively, this raises the possibility that progressive telomere shortening and mitochondrial dysfunction with aging may connectively exacerbate stem cell and tissue aging.

Further recent connections between telomere biology and metabolism have been made in mouse deficiency models linking the deficit in the shelterin component RAP1 to the development of obesity and insulin resistance with age.[68,69]

The precise mechanisms that govern these biological observations linking telomere, mitochondrial biology, and metabolism will undoubtedly be the subject of further investigation to elucidate the molecular pathways and cause and effect relationships that may be at play and ultimately result in their association with aging phenotypes.

3. TELOMERES DYNAMICS IN EARLY EMBRYONIC DEVELOPMENT, STEM CELLS, AND INFANCY

3.1. Gametes, embryogenesis, and stem cells

The telomere set point, or telomere length at birth, is determined by heritable contributions from sperm and oocyte, respectively.[70,71] Telomere length is actively elongated during human spermatogenesis throughout reproductive life, with longer telomeres observed with increasing age.[72] By contrast, primary oocytes develop early in life, arrest, and are subjected to very few divisions during maturation to ovulation, which limits replication-mediated telomere shortening.[73] Telomeres elongate in the early cleavage embryos of mice, which is proposed to occur through a recombination mechanism.[74] The telomerase pathway of telomere lengthening has been shown to be active in early embryogenesis and development and to contribute to the telomere length set point of the individual at birth, effectively "resetting" the telomere length set point for the next generation[75] (Fig. 4.3).

How early in human embryogenesis the telomere length variability range is set and to what extent extragenetic factors may play a role in early human development and influence telomere dynamics have not been formally established (gray box, Fig. 4.2). Embryonic stem cells (ESCs) are derived from the inner cell mass of the blastocyst stage of embryonic development and can be maintained in culture under nondifferentiating conditions indefinitely. ESCs do express telomerase that is required for the maintenance of their proliferative and differentiation potential. Similarly to what is observed

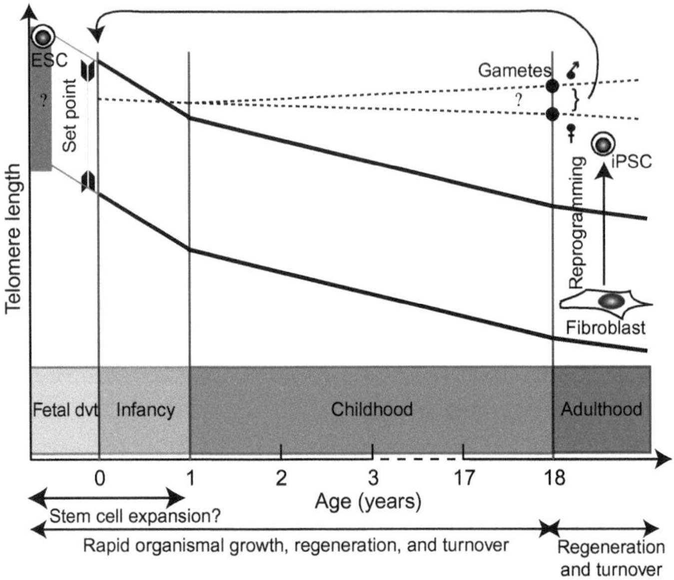

Figure 4.3 Telomere set point, reprogramming and early dynamics. Model of telomere length dynamics in early life and childhood of two hypothetical individuals representing the extent of the normal range (solid bold black lines). Precise dynamics and variables contributing to the telomere length set point (chevrons) in fetal development (dvt) remain unclear (gray box), but a main contribution is from the telomere length of the parental gametes (curved arrow). Germ cells giving rise to gametes retain long telomere lengths (dashed lines) throughout life but show a slight decrease in females due to limited divisions and a slight increase in males due to active lengthening. By comparison, telomere length in somatic cells (solid lines) declines at different rates throughout life and may begin from different set points (chevrons) at birth, in different individuals. The decline in telomere length is most pronounced in the first year of life and hypothesized to follow comparable dynamics as embryonic development (gray lines) when stem cell expansion accompanying rapid phase of growth of the organism occurs. Rapid decline continues during childhood to level off to a cross-sectional steady state rate in adulthood. Adult cells can be reprogrammed to obtain iPSCs that possess ES-like properties achieved in part through resetting telomere length.

between individuals, telomere length heterogeneity is detected between human ESC lines or cultured clones.[76]

The main characteristic of ESCs that makes them attractive for the purpose of regenerative medicine is pluripotency, or the ability to differentiate into all tissue types of the body. Shorter telomere length in ESCs cells has been shown to impact their differentiation potential.[77] Selective expression of the four transcription factors Oct4, Sox2, c-Myc, and Klf4 in adult cells (e.g., such as fibroblasts) leads to reprogramming of these cells into induced pluripotent stem cells or iPSCs.[76,78] The addition of TERT or SV-40 large T antigen to these factors has been shown to support this process and improve efficiency. Telomerase (among other pathways) is activated during this process to confer pluripotency properties to iPSCs that are comparable to those of ESCs. c-Myc and Klf4 have been identified as factors that can bind the TERT gene promoter[79,80] making these factors important regulators of telomere dynamics. Further, the Wnt/β-catenin pathway has recently been connected to TERT expression, supporting a feedforward loop between telomerase signaling and Wnt signaling in the context of high telomerase expression as seen in development and during stem cell renewal.[79,81] Consequently, activating modifications in the regulation or relative expression levels of c-Myc, Klf4, or Wnt can help counter telomere shortening (and may also play a role in mediating pathology particularly in cancer when active or reactivated by a deregulation mechanism; see Section 4 and Chapters 2 and 5).

The reprogramming process once completed achieves significant telomere elongation that effectively reverses and resets the telomere length dynamics of the cell of origin (Fig. 4.2). However, this process has also been shown to be very heterogeneous with an overall low efficiency for reprogramming.[82–84] In addition, reprogramming of a mouse model adult cell type is not equal to what can be achieved by somatic nuclear transfer into an oocyte, where telomere length resetting has been shown to be superior, even in a telomerase-deficient background.[85] This appears to be due in part to differential mitochondrial factors between the recipient oocytes of an adult nucleus or adult cells that undergo reprogramming. Nevertheless, reprogramming and reprogrammed adult cells offer excellent opportunities for disease modeling, intervention testing, and the investigation of regulatory factors that influence telomere dynamics.

More details about the telomere biology of ESCs and iPSCs are addressed in Chapter 3.

3.2. Inherited telomere length and dynamics in infancy

Telomere length has been shown to be in part a heritable genetic trait,[70,71] with paternal age dominance in humans.[86] Cross-sectional studies have demonstrated that leukocyte telomere length decreases during fetal development (as illustrated by measurements in human fetal liver and cord blood[87]) and has a wide range of distribution at birth (as measured in umbilical cord blood samples). The contributions of inherited telomere length from the gametes have been shown to be dominant factors in determining this range; however, early development dynamics are also likely contributors and are currently not well defined. Leukocyte telomere length declines in an age-dependent manner reflecting the high turnover of the hematopoietic tissue and the rate of renewal and differentiation divisions of its stem cells. The telomere length range distribution observed at birth has been shown to remain within a comparable order of magnitude throughout life (Figs. 4.2 and 4.3). It is tempting to speculate that this telomere length range distribution mediated by limited proliferation upon shortening and trimming mechanisms upon excessive lengthening may have evolved to promote optimal genomic stability.

To date, there are still relatively little published data that relate to the longitudinal study of telomere length dynamics in the same human subjects due to the facts that (1) measurement methods have only been available and applied widely to testing individuals for a restricted period of time and (2) suitable human biological samples may not be readily available in existing biobank collections and may need to be collected prospectively, which will be a lengthy process in order to cover the entire human life span. As a consequence, much of the data available are cross-sectional and almost all longitudinal data obtained to date have been collected for adult subjects.[61] Telomere shortening is most rapid during development (comparing fetal liver and cord blood data) and the first year of life,[41,88] due to the fast growth rate the organism experiences during these phases. Longitudinal data in the first year of life from nonhuman primates support a rapid decline in leukocyte telomere length during this period.[89] Respective telomere dynamics may differ during this period between tissues and be dependent on the relative proliferative burden of the tissue and permissive and selective expression of telomerase. Tissue-specific telomere length differences would be predicted to arise during these developmental and early growth stages of the organism.

4. TELOMERE DYNAMICS AND AGING-RELATED DISORDERS

4.1. Telomere deficiency syndromes

Accelerated aging disorders are defined by disease phenotypes that resemble those observed with advancing age, albeit in younger individuals. Telomere deficiency syndromes or telomeropathies are defined as accelerated aging disorders and were first associated with telomere length deficiency through the study of the rare disorder dyskeratosis congenita (DC). DC is characterized by a set of mucocutaneous features associated with bone marrow failure.[90] The discovery of germline DKC1 (encoding dyskerin, an essential minimal component of the telomerase holoenzyme) mutations in X-linked DC implicated telomerase and telomere dynamics in the disease pathology[91] and mediated a failure of tissue homeostasis due to stem cell loss and renewal deficiency.[41,92] This affects the organism as a whole but tends to be most severe in patients with germline mutations and at a younger age in tissues with fast turnover such as the hematopoietic and immune systems (bone marrow failure and immune deficiency often present as the most significant morbidity). The lung, liver, and gut are also organs that are significantly impacted by telomere deficiency over a lifetime and can give rise to disease with loss of function (see Refs. 92–94 and Chapters 5–8). These disease manifestations are consistent with the hypothesis of a loss of tissue stem cells due to telomere shortening that is thought to occur progressively albeit with much slower dynamics during normal aging (Fig. 4.4).

Telomere dynamics are profoundly affected in DC, and the observation of very short telomeres in patients has now been adopted in clinical practice[92] as one of the several diagnostic criteria to help inform treatment decisions. Longitudinal telomere length measurements in DC patients suggest accelerated leukocyte telomere shortening occurs with age in comparison with the cross-sectional data obtained from healthy individuals.[95] Remarkably, telomere dynamics are disrupted due to deficiency or expression level change in a single factor. Most commonly, the deficiency results from a haploinsufficiency that would reduce this level to at most 50% of wild type, is not sufficient to maintain telomere homeostasis, and gives rise to disease in DC. Telomere deficiency diseases and associated causal germline mutations have highlighted what may be the dominant determinants of telomere length regulation and maintenance in human cells over a lifetime. To date, these include the three essential components of the telomerase holoenzyme (*DKC1*, *TERT*, and *TERC*), shuttling factor *TCAB1*, snRNP components *NHP2* and *NOP10*, the shelterin component *TINF2*, and the regulator of telomere length helicase *RTEL1* (see Fig. 4.4 and Chapter 4). The study of individuals with DC has demonstrated that mutations in any of these factors have a significant effect on leukocyte telomere length where the majority of cells are measured with median lengths at the lower limit compatible with cell viability. Further, the inheritance of shorter telomere lengths from an affected parent together with the mutation leads to genetic anticipation[96] and has been shown in a mouse model to lead to degenerative defects

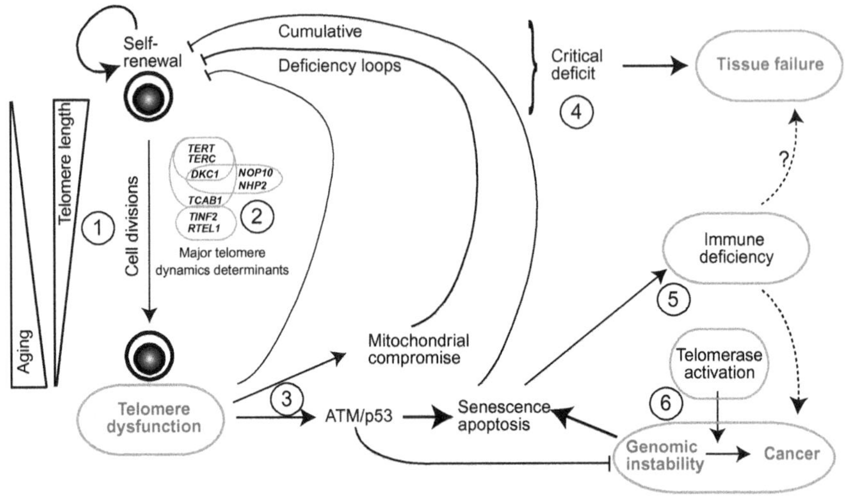

Figure 4.4 Model: consequences of telomere shortening with aging. Telomere length declines with aging and with cell divisions (1), influenced by major determinants of telomere dynamics, and mutations, which have been shown to lead to telomere homeostasis collapse (2). Ensuing telomere dysfunction can prevent adequate stem cell function and differentiation, can trigger the ATM/p53 checkpoint pathway, and can lead to mitochondrial compromise (3). These events have an increasing chance of occurring with aging and collectively accumulate following further telomere shortening. As a result, critical stem cell deficit can lead to tissue failure syndromes(4) and loss of immune cells can lead to deficiency (with the associated increased risk of infection, defective clearance of senescent cells, and cancer) (5). Cells that harbor telomere dysfunction and are able to bypass the ATM/p53 checkpoint can progress through a period of genomic instability and reactivate telomerase to malignant transformation (6).

associated with aging mediated by short telomeres only, in the absence of inheritance of the mutation.[62]

Perhaps surprisingly, despite predictions that shelterin components would be good candidates, no mutations in other shelterin component besides *TINF2* have been identified to date in DC individuals with disease of unknown etiology. This may suggest that point mutations causing such disruptions in telomere function and dynamics mediated by shelterin telomere protection could be incompatible with embryonic development, which is supported by data from knockout mouse models (reviewed in Ref. 97).

Defective telomere dynamics as seen in DC are compatible with the concept that this would progress with stem cell self-renewal or differentiation division and with aging leading to genomic instability.

4.2. Genomic instability and cancer

The connection between telomere biology and malignancy was first proposed when telomerase expression was found to be a hallmark of human cancer: telomerase expression or renewed expression and activity can be detected in 90% of tumor samples (reviewed in Refs. 4,98; the remaining 10% maintain telomeres through the recombination-based alternative lengthening of telomeres mechanism).

DC patients have been shown to be susceptible to genomic instability with an increased lifetime risk of developing cancer.[99] Interestingly, several *TERT* variants are associated with the risk of developing ovarian or breast cancer, thought to occur through silencing or a truncated splice variant.[49] In a separate study, telomere length was shown to mediate genetic anticipation in familial breast cancer,[100] which suggests that both defective telomerase expression and shorter telomere length may be risk factors for these diseases. Progressive earlier age of cancer onset is seen in Li–Fraumeni syndrome (LFS), a cancer predisposition syndrome associated with germ line *TP53* mutations, and is associated with decrease in telomere length[101] providing a link between p53 and telomere biology.

The increase in cancer occurrence may seem paradoxical since the stem cell deficit leading to tissue failure in DC is mediated by telomere dysfunction due to a lack of telomerase, which would ordinarily result in the activation of the replicative senescence checkpoint. However, the strong homeostatic tissue renewal requirements of fast turnover tissues select for cells that can bypass the ATM/p53 checkpoint (defective in LFS). The continued cell division of these genetically unstable cells with telomere dysfunction leads to chromosome breakage–fusion–bridge cycles (crisis) from which the great majority of cells will undergo apoptosis. Abnormal malignant clones with upregulated telomerase expression may emerge, mediated by *TERT* copy number variations, gene amplification[102,103], or modification of any of the regulators of *TERT* expression described in Section 1 (e.g., through the upregulation of c-Myc). These cancer cells have regained telomere capping capability or have gained an unlimited proliferation potential through the continued expression of telomerase. Ectopic hTERT expression permits postsenescent cells to proliferate beyond crisis, in a process that could to be independent of catalytic activity.[104] These cancer cells typically display short telomeres that remain stable over time, may have defective capping capacity, or may harbor somatic mutations in telomere capping components that may be driving further genomic instability and tumor evolution.

5. PERSPECTIVE

Continued studies of complex orchestration of the determinants and modifiers of telomere dynamics will continue to inform our knowledge of telomere biology in relation to health, disease, and aging. Progress in this area will help determine which molecules or pathways may be affectively targeted to treat specific deficiencies and cancer and may help improve the health status in aging individuals.

ACKNOWLEDGMENTS

I am grateful to Dr. Peter Lansdorp (European Research Institute for the Biology of Ageing, University of Groningen, and Terry Fox Laboratory, B.C. Cancer Agency, Vancouver, Canada) and Dr. Ester Falconer (Terry Fox Laboratory) for their careful reading of the manuscript and insightful comments.

Work in the Lansdorp laboratory in Vancouver (Terry Fox Laboratory) is supported by grants from the Canadian Institutes of Health Research (RMF-92093), the US National Institutes of Health Research (R01GM094146), the Canadian Cancer Society, and the Terry Fox Foundation (Grants 018006 and 105265).

Conflict of interest disclosure: G. A. holds part-time employment at Repeat Diagnostics Inc., a company specializing in leukocyte telomere length measurements using Flow-FISH.

REFERENCES

1. Morin GB. The human telomere terminal transferase enzyme is a ribonucleoprotein that synthesizes TTAGGG repeats. *Cell*. 1989;59(3):521–529.
2. Moyzis RK, Buckingham JM, Cram LS, et al. A highly conserved repetitive DNA sequence, (TTAGGG)n, present at the telomeres of human chromosomes. *Proc Natl Acad Sci USA*. 1988;85(18):6622–6626.
3. Greider CW, Blackburn EH. Identification of a specific telomere terminal transferase activity in Tetrahymena extracts. *Cell*. 1985;43(2 Pt 1):405–413.
4. Hanahan D, Weinberg RA. The hallmarks of cancer. *Cell*. 2000;100(1):57–70.
5. Hanahan D, Weinberg RA. Hallmarks of cancer: the next generation. *Cell*. 2011;144(5):646–674.
6. Lopez-Otin C, Blasco MA, Partridge L, Serrano M, Kroemer G. The hallmarks of aging. *Cell*. 2013;153(6):1194–1217.
7. de Lange T. Shelterin: the protein complex that shapes and safeguards human telomeres. *Genes Dev*. 2005;19(18):2100–2110.
8. Sfeir A, de Lange T. Removal of shelterin reveals the telomere end-protection problem. *Science*. 2012;336(6081):593–597.
9. Songyang Z, Liu D. Inside the mammalian telomere interactome: regulation and regulatory activities of telomeres. *Crit Rev Eukaryot Gene Expr*. 2006;16(2):103–118.
10. Amiard S, Doudeau M, Pinte S, et al. A topological mechanism for TRF2-enhanced strand invasion. *Nat Struct Mol Biol*. 2007;14(2):147–154.
11. Bianchi A, Smith S, Chong L, Elias P, de Lange T. TRF1 is a dimer and bends telomeric DNA. *EMBO J*. 1997;16(7):1785–1794.
12. Griffith JD, Comeau L, Rosenfield S, et al. Mammalian telomeres end in a large duplex loop. *Cell*. 1999;97(4):503–514.

13. Olovnikov AM. A theory of marginotomy. The incomplete copying of template margin in enzymic synthesis of polynucleotides and biological significance of the phenomenon. *J Theor Biol.* 1973;41(1):181–190.
14. Wu P, Takai H, de Lange T. Telomeric 3′ overhangs derive from resection by Exo1 and Apollo and fill-in by POT1b-associated CST. *Cell.* 2012;150(1):39–52.
15. Miyake Y, Nakamura M, Nabetani A, et al. RPA-like mammalian Ctc1-Stn1-Ten1 complex binds to single-stranded DNA and protects telomeres independently of the Pot1 pathway. *Mol Cell.* 2009;36(2):193–206.
16. Surovtseva YV, Churikov D, Boltz KA, et al. Conserved telomere maintenance component 1 interacts with STN1 and maintains chromosome ends in higher eukaryotes. *Mol Cell.* 2009;36(2):207–218.
17. Verdun RE, Crabbe L, Haggblom C, Karlseder J. Functional human telomeres are recognized as DNA damage in G2 of the cell cycle. *Mol Cell.* 2005;20(4):551–561.
18. Pickett H, Cesare A, Johnston R, Neumann A, Reddel R. Control of telomere length by a trimming mechanism that involves generation of t-circles EMBO open. *EMBO J.* 2009;28(7):799–809.
19. Uringa EJ, Lisaingo K, Pickett HA, et al. RTEL1 contributes to DNA replication and repair and telomere maintenance. *Mol Biol Cell.* 2012;23(14):2782–2792.
20. Vannier JB, Pavicic-Kaltenbrunner V, Petalcorin MI, Ding H, Boulton SJ. RTEL1 dismantles T loops and counteracts telomeric G4-DNA to maintain telomere integrity. *Cell.* 2012;149(4):795–806.
21. Harrington L, McPhail T, Mar V, et al. A mammalian telomerase-associated protein. *Science.* 1997;275(5302):973–977.
22. Nakamura TM, Morin GB, Chapman KB, et al. Telomerase catalytic subunit homologs from fission yeast and human. *Science.* 1997;277(5328):955–959.
23. Feng J, Funk WD, Wang SS, et al. The RNA component of human telomerase. *Science.* 1995;269(5228):1236–1241.
24. Mitchell JR, Wood E, Collins K. A telomerase component is defective in the human disease dyskeratosis congenita. *Nature.* 1999;402(6761):551–555.
25. Wong JM, Collins K. Telomerase RNA level limits telomere maintenance in X-linked dyskeratosis congenita. *Genes Dev.* 2006;20(20):2848–2858.
26. Cohen SB, Graham ME, Lovrecz GO, Bache N, Robinson PJ, Reddel RR. Protein composition of catalytically active human telomerase from immortal cells. *Science.* 2007;315(5820):1850–1853.
27. Sauerwald A, Sandin S, Cristofari G, Scheres SH, Lingner J, Rhodes D. Structure of active dimeric human telomerase. *Nat Struct Mol Biol.* 2013;20(4):454–460.
28. Stern JL, Zyner KG, Pickett HA, Cohen SB, Bryan TM. Telomerase recruitment requires both TCAB1 and Cajal bodies independently. *Mol Cell Biol.* 2012;32(13): 2384–2395.
29. Venteicher AS, Abreu EB, Meng Z, et al. A human telomerase holoenzyme protein required for Cajal body localization and telomere synthesis. *Science.* 2009;323(5914): 644–648.
30. Kappei D, Butter F, Benda C, et al. HOT1 is a mammalian direct telomere repeat-binding protein contributing to telomerase recruitment. *EMBO J.* 2013;32(12):1681–1701.
31. Wang F, Podell ER, Zaug AJ, et al. The POT1-TPP1 telomere complex is a telomerase processivity factor. *Nature.* 2007;445(7127):506–510.
32. Xin H, Liu D, Wan M, et al. TPP1 is a homologue of ciliate TEBP-beta and interacts with POT1 to recruit telomerase. *Nature.* 2007;445(7127):559–562.
33. Chen LY, Redon S, Lingner J. The human CST complex is a terminator of telomerase activity. *Nature.* 2012;488(7412):540–544.
34. Azzalin CM, Reichenbach P, Khoriauli L, Giulotto E, Lingner J. Telomeric repeat containing RNA and RNA surveillance factors at mammalian chromosome ends. *Science.* 2007;318(5851):798–801.

35. Farnung BO, Brun CM, Arora R, Lorenzi LE, Azzalin CM. Telomerase efficiently elongates highly transcribing telomeres in human cancer cells. *PLoS One*. 2012;7(4): e35714.
36. Pfeiffer V, Lingner J. TERRA promotes telomere shortening through exonuclease 1-mediated resection of chromosome ends. *PLoS Genet*. 2012;8(6):e1002747.
37. Schoeftner S, Blasco MA. Chromatin regulation and non-coding RNAs at mammalian telomeres. *Semin Cell Dev Biol*. 2010;21(2):186–193.
38. Houghtaling BR, Canudas S, Smith S. A role for sister telomere cohesion in telomere elongation by telomerase. *Cell Cycle*. 2012;11(1):19–25.
39. Cifuentes-Rojas C, Shippen DE. Telomerase regulation. *Mutat Res*. 2012;730(1–2): 20–27.
40. Aubert G, Hills M, Lansdorp PM. Telomere length measurement-caveats and a critical assessment of the available technologies and tools. *Mutat Res*. 2012;730(1–2):59–67.
41. Aubert G, Baerlocher GM, Vulto I, Poon SS, Lansdorp PM. Collapse of telomere homeostasis in hematopoietic cells caused by heterozygous mutations in telomerase genes. *PLoS Genet*. 2012;8(5):e1002696. http://dx.doi.org/10.1371/journal.pgen.1002696.
42. Kimura M, Gazitt Y, Cao X, Zhao X, Lansdorp PM, Aviv A. Synchrony of telomere length among hematopoietic cells. *Exp Hematol*. 2010;38(10):854–859.
43. Okuda K, Bardeguez A, Gardner JP, et al. Telomere length in the newborn. *Pediatr Res*. 2002;52(3):377–381.
44. Gadalla SM, Cawthon R, Giri N, Alter BP, Savage SA. Telomere length in blood, buccal cells, and fibroblasts from patients with inherited bone marrow failure syndromes. *Aging (Albany NY)*. 2010;2(11):867–874.
45. Kannan N, Huda N, Tu L, et al. The luminal progenitor compartment of the normal human mammary gland constitutes a unique site of telomere dysfunction. *Stem Cell Rep*. 2013;1(1):28–37.
46. Brummendorf TH, Rufer N, Baerlocher GM, Roosnek E, Lansdorp PM. Limited telomere shortening in hematopoietic stem cells after transplantation. *Ann N Y Acad Sci*. 2001;938:1–7, discussion 7–8.
47. Buchkovich KJ, Greider CW. Telomerase regulation during entry into the cell cycle in normal human T cells. *Mol Biol Cell*. 1996;7(9):1443–1454.
48. Ziegler P, Schrezenmeier H, Akkad J, et al. Telomere elongation and clinical response to androgen treatment in a patient with aplastic anemia and a heterozygous hTERT gene mutation. *Ann Hematol*. 2012;91(7):1115–1120.
49. Bojesen SE, Pooley KA, Johnatty SE, et al. Multiple independent variants at the TERT locus are associated with telomere length and risks of breast and ovarian cancer. *Nat Genet*. 2013;45(4):371–384, 384e371–372.
50. Calado RT, Regal JA, Hills M, et al. Constitutional hypomorphic telomerase mutations in patients with acute myeloid leukemia. *Proc Natl Acad Sci USA*. 2009;106(4):1187–1192.
51. Hayflick L. The limited in vitro lifetime of human diploid cell strains. *Exp Cell Res*. 1965;37:614–636.
52. Hayflick L. A brief history of the mortality and immortality of cultured cells. *Keio J Med*. 1998;47(3):174–182.
53. Counter CM, Avilion AA, LeFeuvre CE, et al. Telomere shortening associated with chromosome instability is arrested in immortal cells which express telomerase activity. *EMBO J*. 1992;11(5):1921–1929.
54. Harley CB, Futcher AB, Greider CW. Telomeres shorten during ageing of human fibroblasts. *Nature*. 1990;345(6274):458–460.
55. Hastie ND, Dempster M, Dunlop MG, Thompson AM, Green DK, Allshire RC. Telomere reduction in human colorectal carcinoma and with ageing. *Nature*. 1990;346(6287):866–868.

56. Hayflick L. Living forever and dying in the attempt. *Exp Gerontol.* 2003;38(11–12): 1231–1241.
57. Daniali L, Benetos A, Susser E, et al. Telomeres shorten at equivalent rates in somatic tissues of adults. *Nat Commun.* 2013;4:1597.
58. Artandi SE, DePinho RA. A critical role for telomeres in suppressing and facilitating carcinogenesis. *Curr Opin Genet Dev.* 2000;10(1):39–46.
59. Kaul Z, Cesare AJ, Huschtscha LI, Neumann AA, Reddel RR. Five dysfunctional telomeres predict onset of senescence in human cells. *EMBO Rep.* 2012;13(1):52–59.
60. Campisi J, Andersen JK, Kapahi P, Melov S. Cellular senescence: a link between cancer and age-related degenerative disease? *Semin Cancer Biol.* 2011;21(6):354–359.
61. Steenstrup T, Hjelmborg JV, Kark JD, Christensen K, Aviv A. The telomere lengthening conundrum—artifact or biology? *Nucleic Acids Res.* 2013;41(13):e131.
62. Armanios M, Alder JK, Parry EM, Karim B, Strong MA, Greider CW. Short telomeres are sufficient to cause the degenerative defects associated with aging. *Am J Hum Genet.* 2009;85(6):823–832.
63. Begus-Nahrmann Y, Hartmann D, Kraus J, et al. Transient telomere dysfunction induces chromosomal instability and promotes carcinogenesis. *J Clin Invest.* 2012;122(6):2283–2288.
64. Sahin E, Colla S, Liesa M, et al. Telomere dysfunction induces metabolic and mitochondrial compromise. *Nature.* 2011;470(7334):359–365.
65. Edgar D, Trifunovic A. The mtDNA mutator mouse: dissecting mitochondrial involvement in aging. *Aging (Albany, NY).* 2009;1(12):1028–1032.
66. Bratic A, Larsson NG. The role of mitochondria in aging. *J Clin Invest.* 2013;123(3):951–957.
67. Scheibye-Knudsen M, Scheibye-Alsing K, Canugovi C, Croteau DL, Bohr VA. A novel diagnostic tool reveals mitochondrial pathology in human diseases and aging. *Aging (Albany, NY).* 2013;5(3):192–208.
68. Martinez P, Gomez-Lopez G, Garcia F, et al. RAP1 protects from obesity through its extratelomeric role regulating gene expression. *Cell Rep.* 2013;3(6):2059–2074.
69. Yeung F, Ramirez CM, Mateos-Gomez PA, et al. Nontelomeric role for Rap1 in regulating metabolism and protecting against obesity. *Cell Rep.* 2013;3(6):1847–1856.
70. Graakjaer J, Der-Sarkissian H, Schmitz A, et al. Allele-specific relative telomere lengths are inherited. *Hum Genet.* 2006;119(3):344–350.
71. Slagboom PE, Droog S, Boomsma DI. Genetic determination of telomere size in humans: a twin study of three age groups. *Am J Hum Genet.* 1994;55(5):876–882.
72. Kimura M, Cherkas LF, Kato BS, et al. Offspring's leukocyte telomere length, paternal age, and telomere elongation in sperm. *PLoS Genet.* 2008;4(2):e37.
73. Turner S, Hartshorne GM. Telomere lengths in human pronuclei, oocytes and spermatozoa. *Mol Hum Reprod.* 2013;19(8):510–518.
74. Liu L, Bailey SM, Okuka M, et al. Telomere lengthening early in development. *Nat Cell Biol.* 2007;9(12):1436–1441.
75. Chiang YJ, Calado RT, Hathcock KS, Lansdorp PM, Young NS, Hodes RJ. Telomere length is inherited with resetting of the telomere set-point. *Proc Natl Acad Sci USA.* 2010;107(22):10148–10153.
76. Amit M, Carpenter MK, Inokuma MS, et al. Clonally derived human embryonic stem cell lines maintain pluripotency and proliferative potential for prolonged periods of culture. *Dev Biol.* 2000;227(2):271–278.
77. Pucci F, Gardano L, Harrington L. Short telomeres in ESCs lead to unstable differentiation. *Cell Stem Cell.* 2013;12(4):479–486.
78. Park IH, Zhao R, West JA, et al. Reprogramming of human somatic cells to pluripotency with defined factors. *Nature.* 2008;451(7175):141–146.

79. Hoffmeyer K, Raggioli A, Rudloff S, et al. Wnt/beta-catenin signaling regulates telomerase in stem cells and cancer cells. *Science*. 2012;336(6088):1549–1554.
80. Wu KJ, Grandori C, Amacker M, et al. Direct activation of TERT transcription by c-MYC. *Nat Genet*. 1999;21(2):220–224.
81. Diala I, Wagner N, Magdinier F, et al. Telomere protection and TRF2 expression are enhanced by the canonical Wnt signalling pathway. *EMBO Rep*. 2013;14(4):356–363.
82. Agarwal S, Loh YH, McLoughlin EM, et al. Telomere elongation in induced pluripotent stem cells from dyskeratosis congenita patients. *Nature*. 2010;464(7286):292–296.
83. Liang G, Zhang Y. Genetic and epigenetic variations in iPSCs: potential causes and implications for application. *Cell Stem Cell*. 2013;13(2):149–159.
84. Muller LU, Daley GQ, Williams DA. Upping the ante: recent advances in direct reprogramming. *Mol Ther*. 2009;17(6):947–953.
85. Le R, Kou Z, Jiang Y, et al. Enhanced telomere rejuvenation in pluripotent cells reprogrammed via nuclear transfer relative to induced pluripotent stem cells. *Cell Stem Cell*. 2014;14(1):27–39.
86. Njajou OT, Cawthon RM, Damcott CM, et al. Telomere length is paternally inherited and is associated with parental lifespan. *Proc Natl Acad Sci USA*. 2007;104(29):12135–12139.
87. Vaziri H, Dragowska W, Allsopp RC, Thomas TE, Harley CB, Lansdorp PM. Evidence for a mitotic clock in human hematopoietic stem cells: loss of telomeric DNA with age. *Proc Natl Acad Sci USA*. 1994;91(21):9857–9860.
88. Rufer N, Brummendorf TH, Kolvraa S, et al. Telomere fluorescence measurements in granulocytes and T lymphocyte subsets point to a high turnover of hematopoietic stem cells and memory T cells in early childhood. *J Exp Med*. 1999;190(2):157–167.
89. Baerlocher GM, Rice K, Vulto I, Lansdorp PM. Longitudinal data on telomere length in leukocytes from newborn baboons support a marked drop in stem cell turnover around 1 year of age. *Aging Cell*. 2007;6(1):121–123.
90. Savage SA, Dokal I, Armanios M, et al. Dyskeratosis congenita: the first NIH clinical research workshop. *Pediatr Blood Cancer*. 2009;53(3):520–523.
91. Heiss NS, Knight SW, Vulliamy TJ, et al. X-linked dyskeratosis congenita is caused by mutations in a highly conserved gene with putative nucleolar functions. *Nat Genet*. 1998;19(1):32–38.
92. Alter BP, Baerlocher GM, Savage SA, et al. Very short telomere length by flow fluorescence in situ hybridization identifies patients with dyskeratosis congenita. *Blood*. 2007;110(5):1439–1447.
93. Armanios MY, Chen JJ, Cogan JD, et al. Telomerase mutations in families with idiopathic pulmonary fibrosis. *N Engl J Med*. 2007;356(13):1317–1326.
94. Calado RT, Regal JA, Kleiner DE, et al. A spectrum of severe familial liver disorders associate with telomerase mutations. *PLoS One*. 2009;4(11):e7926.
95. Alter BP, Rosenberg PS, Giri N, Baerlocher GM, Lansdorp PM, Savage SA. Telomere length is associated with disease severity and declines with age in dyskeratosis congenita. *Haematologica*. 2012;97(3):353–359.
96. Vulliamy T, Marrone A, Szydlo R, Walne A, Mason PJ, Dokal I. Disease anticipation is associated with progressive telomere shortening in families with dyskeratosis congenita due to mutations in TERC. *Nat Genet*. 2004;36(5):447–449.
97. Chan SS, Chang S. Defending the end zone: studying the players involved in protecting chromosome ends. *FEBS Lett*. 2010;584(17):3773–3778.
98. Shay JW, Roninson IB. Hallmarks of senescence in carcinogenesis and cancer therapy. *Oncogene*. 2004;23(16):2919–2933.
99. Alter BP, Giri N, Savage SA, Rosenberg PS. Cancer in dyskeratosis congenita. *Blood*. 2009;113(26):6549–6557.

100. Martinez-Delgado B, Yanowsky K, Inglada-Perez L, et al. Genetic anticipation is associated with telomere shortening in hereditary breast cancer. *PLoS Genet.* 2011;7(7):e1002182.
101. Tabori U, Nanda S, Druker H, Lees J, Malkin D. Younger age of cancer initiation is associated with shorter telomere length in Li–Fraumeni syndrome. *Cancer Res.* 2007;67(4):1415–1418.
102. Cao Y, Bryan TM, Reddel RR. Increased copy number of the TERT and TERC telomerase subunit genes in cancer cells. *Cancer Sci.* 2008;99(6):1092–1099.
103. Weir BA, Woo MS, Getz G, et al. Characterizing the cancer genome in lung adenocarcinoma. *Nature.* 2007;450(7171):893–898.
104. Counter CM, Hahn WC, Wei W, et al. Dissociation among in vitro telomerase activity, telomere maintenance, and cellular immortalization. *Proc Natl Acad Sci USA.* 1998;95(25):14723–14728.

CHAPTER FIVE

Telomeres in Molecular Epidemiology Studies

Clara Bodelon, Sharon A. Savage, Shahinaz M. Gadalla
Division of Cancer Epidemiology and Genetics, National Cancer Institute, National Institutes of Health, Bethesda, Maryland, USA

Contents

Abstract

Telomeres are long nucleotide repeats and protein complexes at the ends of chromosomes that are essential for maintaining chromosomal stability. They shorten with each cell division, and therefore, telomere length is a marker for cellular aging and senescence. Epidemiological research of telomeres investigates the role that these genetic structures have in disease risk and mortality in human populations. This chapter provides an overview of the current telomere epidemiology research and discusses approaches taken in these investigations. We also highlight important methodological considerations that may affect data interpretation.

1. INTRODUCTION

Telomeres have been an attractive biological marker for longevity and health since the discovery that they shorten with normal cell division.[1] Telomere studies in humans have ranged from understanding telomeric structure, function, and regulatory mechanisms to evaluating their role in aging, disease risk, and mortality. At the end of 2013, the search term

Progress in Molecular Biology and Translational Science
Volume 125, ISSN 1877-1173
http://dx.doi.org/10.1016/B978-0-12-397898-1.00005-0

"telomeres" in PubMed returned more than 17,000 publications. The first,[2] published in 1950, reported the existence of a structure that was first hypothesized by Hermann Muller in the 1930s to cap the end of the chromosome. Since then, the number of telomere-related publications has grown rapidly (Fig. 5.1). Telomere studies have been largely conducted in the fields of cancer and aging as noted by the number of publications in those fields. As of early 2014, the search term "telomeres and cancer" returned 5586 publications and "telomeres and aging" returned 3340 publications.

Epidemiology is derived from three Greek words: *epi*, meaning "among"; *demos*, meaning "people"; and *logos* meaning "study." In total, they mean "the study of what is among the people," with a focus on health-related events. The science of epidemiology further expanded from describing distribution and patterns to identifying etiological factors and evaluating effects of diseases in a defined population.

The epidemiological literature of telomeres over the past decade features studies focused on (1) describing telomere length (TL) variation by demographic characteristics such as age,[3] gender, and race[4]; (2) evaluating the association between rare and common inherited telomeric defects and disease risk or severity[5–7]; (3) evaluating the effect of environmental exposures and lifestyle factors on TL (e.g., stress,[8,9] exercise,[10] smoking,[11,12] and

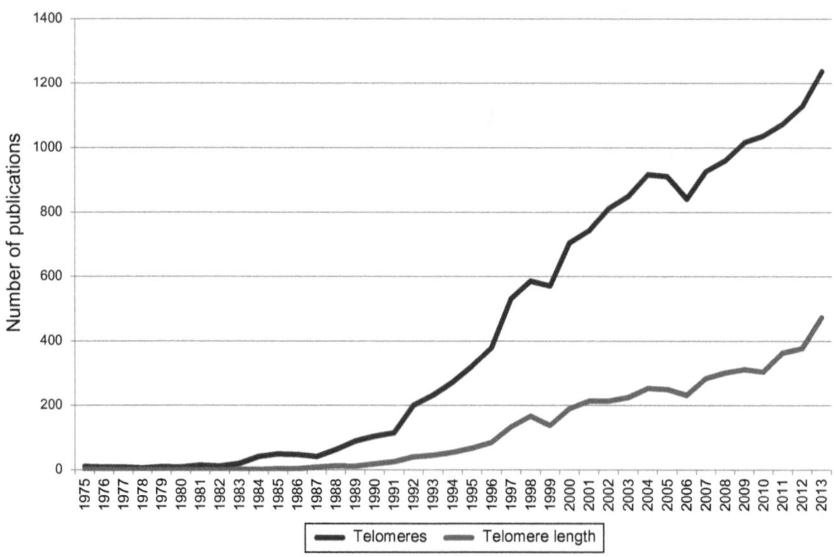

Figure 5.1 Number of telomere-related PubMed-indexed publications and the observed rising trend overtime.

certain chemicals[13]); and (4) determining the role telomeres may play in certain outcomes such as mortality[14] or age-related diseases such as cancer,[15] cardiovascular diseases,[16] and other diseases.[17] Few behavioral intervention studies have been implemented to date; they include a study evaluating the role of meditation training on telomerase activity[18] and a study of the role of a lifestyle intervention on TL.[19]

In this chapter, we discuss challenges related to the study design in telomere epidemiology research, limitations related to telomere measurement methods, and surrogate tissue use in most studies. Table 5.1 presents the definitions of some epidemiological terms[20,21] introduced in this chapter.

Table 5.1 Definitions of some epidemiological and statistical terms

Term	Definition
Confounding	The distortion of the effect of an exposure on an outcome due to the association of another factor (confounder) with the exposure and the outcome
Bias	Systematic deviation of results or inferences from truth
Temporal relationship	Exposure always precedes the outcome
Biological plausibility	The association is coherent with firmly established knowledge on pathobiological processes
Mediator variable	A variable that occurs in the pathway from an exposure to an outcome
Correlation coefficient (r)	A measure of association that indicates the degree to which two variables have a linear relationship It ranges between −1 and 1; $r=0$ means no correlation
Coefficient of variation (CV)	A measure of the spread of the data defined as the ratio of the standard deviation to the mean. Usually expressed as a percentage
Odds ratio (OR)	The ratio of the odds of exposure among the cases to the odds of exposure among controls; it is commonly used in case-control studies
Hazard ratio (HR)	The ratio of the hazard rates comparing exposed with unexposed groups. Hazard rate is the number of new cases developing per unit person-time at risk
Confidence interval (CI)	A range of values about a point estimate that indicates the degree of statistical precision that describes the estimate A wider interval indicates less precision

2. TL IN ACCESSIBLE TISSUE AS A SURROGATE FOR THE DISEASE SITE

Many epidemiological studies measure TL in peripheral blood leukocytes. While leukocyte TL may be the specific biomarker of interest in some studies, in others, it may be a surrogate for a specific tissue. Data comparing TL in peripheral blood leukocytes with TL in different somatic tissues from the same individual are limited. The use of leukocyte TL as a surrogate for the tissue of interest may be affected by the tissue-specific proliferation rate and the susceptibility to environmental factors affecting TL regulation in that specific tissue. Leukocytes are a heterogeneous cell population, formed mainly by lymphocytes and granulocytes. Notably, lymphocytes have shorter telomeres than granulocytes.[22]

Studies comparing TL in different tissues from normal individuals have found differences in absolute TL between tissues, but strong correlations are present. For instance, in a study measuring TL in leukocytes, skeletal muscle, skin, and subcutaneous fat from the same 87 adult individuals, the authors found high degrees of correlation within individuals but noted that leukocytes had the shortest telomeres.[23] Similar findings have been reported when comparing TL in leukocytes, skin, and joint tissues in a small study of elderly patients with hip fractures.[24] A Japanese study measured TL in cerebral cortex, kidney, liver, and myocardium in over 100 autopsies from individuals of different ages, including neonates through centenarians, and also found high intraindividual correlations between different tissue TL measurements[25]; the myocardial tissue had the longest TL, while the liver and renal cortex had the shortest. Finally, a study compared average leukocyte TL and buccal TL in young (18–26 years of age) and old (66–75 years of age) adults.[26] The authors found that leukocyte TL was significantly shorter than buccal TL in both age groups. As expected, leukocyte TL was significantly shorter in the older group compared with that of the younger group, but surprisingly, there were no differences in buccal TL between the two age groups.[26] Of note, other factors, such as increased body mass index (BMI) or tobacco smoking, which have been associated with shortening of leukocyte TL,[12] were not accounted for in that study.

Some studies have compared TL measured in different tissues from individuals with specific illnesses. TL was measured in leukocytes, buccal cells, and fibroblasts in individuals with dyskeratosis congenita (DC), an inherited telomere biology disorder where telomeres are exceedingly short (below the first percentile of normal individuals of the same age).[5,7] The authors reported strong correlations between the TLs in these different types of cells, with the leukocytes having the shortest telomeres.[27] TL has also been measured in somatic tissues from cancer patients, comparing tumor tissue to adjacent normal tissues in most cases. Telomeres were shorter in prostate cancer cells than in adjacent nonmalignant cells.[28] Moreover, shorter TL in prostate cancer-associated stromal cells and larger TL variation in prostate cancer cells were associated with poor outcomes. Telomere shortening has also been reported in the early stages of breast carcinogenesis[29] and was correlated with more aggressive subtypes of breast cancer.[30] Telomeres in tissues near breast tumors were shorter compared with those in distant normal breast tissues[31,32] suggesting a TL field effect in the adjacent nonmalignant breast tissue. The variation in TL in adjacent nonmalignant breast tissues was suggested to be a prognostic factor of breast cancer recurrence.[33] Finally, findings from a study that measured TL in colon tissues from patients diagnosed with ulcerative colitis, a chronic inflammatory disease of the colon, and healthy controls suggested shorter colonic TL in patients than in healthy controls.[34] Notably, telomere shortening has also been observed in adenomas and colorectal cancer tissues as compared with normal adjacent or distal colon tissues.[35,36] The field effect suggested in the breast cancer studies did not appear to be present in a colon cancer and adenoma study.[36]

Overall, there is evidence of strong correlation between leukocyte TL and TL measured in other somatic tissues from the same individuals, which is not surprising, given that TL is highly heritable.[37] It is important to note that the absolute TL differs between tissues and telomeres tend to be shorter in leukocytes than in other evaluated tissues. This suggests that the dynamics of TL are likely to be tissue-specific, reflecting tissue-specific replicative capacity. Future research is required to evaluate the relationship between TL measured in surrogate tissues and TL measured in tissues affected by the disease of interest.

3. CONSIDERATIONS IN OBSERVATIONAL STUDIES OF TL

Association studies of TL and disease have become an important area in molecular and genetic epidemiology. Like other genetic and biological marker studies, telomere epidemiology relies on the common observational study designs, including cross-sectional, case-control, and cohort studies.

Table 5.2 Observational study designs

Study design	Purpose	Characteristics
Cross-sectional	– To assess the prevalence of an exposure or conditions – Sometimes used to evaluate the association between study variables (exposures and outcomes)	– Simple to conduct – Suitable for assessing multiple exposures and outcomes – Information on exposure and outcome is obtained at the same time; cannot assess temporality
Case-control	– To compare odds of exposure (e.g., telomere length) in diseased and not diseased populations	– Retrospective study (recruitment happens after disease develops) – Suitable for rare diseases – Suitable for assessing multiple exposures – Controls need to be representative of the population that gave rise to the cases; otherwise subject to selection bias – Subject to reverse causation – Subject to recall bias
Cohort	– To compare disease incidence over time in exposed *versus* nonexposed individuals	– Prospective study (participants are free from outcome at the start of the study) – Suitable for assessing multiple outcomes – Long follow-up times and large samples are needed for rare outcomes; it can be costly – Subject to bias from nonresponse or losses to follow-up

Table 5.2 summarizes the purpose and characteristics of each design, which should be carefully considered during study design and data interpretation.

Cross-sectional studies in telomere epidemiology have been used to assess the relationship between TL and demographic characteristics, such as behavioral lifestyles or environmental exposures. For example, the cross-sectional study design has been used to evaluate the relationship between TL and chronological age (reviewed in Ref. 38) and its distribution by sex and race.[4] Overall, studies have shown that older age,[39] White race,[4] and male gender[4,40] are associated with shorter telomeres, but the strength of association varied by the participants' age and method of TL measurement.[41] The association between TL and other factors, such as smoking, physical activity, and alcohol use, has been inconsistent.[11,12,42] The inconsistencies found in the published studies may be related to the fact that TL at the time of the study may not reflect the current state of an exposure, as measured in a cross-sectional study.[43] Cross-sectional studies are often highly feasible but limited by their inability to determine a temporal relationship between TL and factors of interest (i.e., outcome may have preceded TL changes in this situation). Despite this limitation, cross-sectional studies are a valuable tool for hypothesis generation.

Cross-sectional studies are useful in evaluating the relationship between genetic variation in telomere genes and disease risk since an individual germline genome does not change over time. Single-nucleotide polymorphisms (SNPs) in the *TERT* gene locus have been associated with peripheral arterial disease[44] and several cancers[45,46]; variation in the *OBFC1* gene predicted cardiovascular mortality in women[47]; and variation in both *TERT* and *TEP1* genes was associated with male infertility.[48] In these situations, investigators suggest that genetic variation can be thought as a surrogate of TL.[49] This is based on a connection between TL and SNPs in several genome-wide association studies that identified associations between genetic variation in some known telomere biology genes (*TERT*, *TERC*, and *OBFC1*) and TL.[45,50,51] However, many more studies are required before making a direct connection between these SNPs and TL.

Case-control and prospective studies are commonly used analytic approaches in evaluating whether TL is a biomarker or risk factor for age-related diseases. The main difference between the two approaches is the timing of TL assessment in relation to the disease of interest. In the case-control design, TL is measured after the disease develops and is compared with TL in individuals who are free of the disease (controls). The selection of the control population is crucial; it should be representative to the source population that gave rise to the cases to avoid possible selection bias. Details about control selection processes can be found elsewhere.[52] In the prospective design (cohort, nested case-control, or case-cohort) studies, TL is measured in disease-free individuals who are followed up overtime, and disease rate is then assessed in relation to baseline TL. Because classic cohort studies are expensive and usually require long follow-up time, prospective nested case-control and case-cohort studies are valuable because they include advantages present in both case-control and cohort studies.[52]

In this approach, cases and controls are selected from an existing cohort in whom biological samples have been collected in the past and prior to disease development. Controls in the nested case-control design are a random sample from subjects who have not developed the disease at the time of case diagnosis, but are a sample from all cohort participants in the case-cohort design. The biospecimen collection precedes disease development; thus, TL precedes disease diagnosis in these studies.

Association studies of TL and cancer are an example of why study design is crucial in the conduct and interpretation of TL epidemiology studies. Many studies have suggested that short telomeres are associated with cancer risk.[15,53] However, these data were primarily derived from case-control studies, which could be subject to reverse-causation bias. The meta-analysis by Wentzensen and colleagues showed significant differences between findings of TL and cancer associations in case-control studies and those in cohort studies (odds ratio (OR) = 2.90, 95% confidence interval (CI) = 1.75–4.80 in case-control studies vs. OR = 1.16, 95% CI = 0.87–1.54 in prospective studies[15]). This observed difference could be, at least partially, explained by a possible reverse-causation bias in case-control studies; that is, cancer precedes telomere shortening and the observed difference in TL between cases and controls is due to the disease status. This bias is less of a concern in prospective studies. In all prospective studies, it is important to assess the lag time between sample collection and disease development, since chronic diseases such as heart disease or even cancer can develop over a long period of time. If DNA samples are collected close to the date of diagnosis, the disease could already be present, but not diagnosed, and therefore, this relationship will still suffer from possible reverse causation.

Another example on how the choice of the study design can influence the findings comes from epidemiological studies evaluating age-related telomere attrition rate. Most of these studies have been cross-sectional in nature; only few longitudinal studies have been reported. Cross-sectional studies attempt to estimate such attrition rate based on a single measurement of TL in subjects of different ages. They assume that the telomere attrition rate is constant across different ages and is similar in all individuals. The age range of study participants (telomere erosion varies across ages, with faster erosion during childhood), their gender, race distribution, and survival effect may also affect these estimates.[54,55] Longitudinal studies have calculated the rate of TL attrition based on two measurements at a 3- to 10-year time interval and suggest that between 38 and 48 base pairs of telomeres are lost per year.[38] When cross-sectional and longitudinal assessments of TL attrition rate were simultaneously compared, longitudinal TL attrition varied between individuals with 10% of study participants experiencing TL elongation overtime.[56] Whether TL elongation over time is a true phenomenon or an artifact caused by measurement error is yet to be determined.[57] Long follow-up periods, large sample sizes, and multiple TL measurement points are needed to thoroughly understand the age-related decline in TL.

4. EPIDEMIOLOGICAL EVALUATION OF THE ROLE OF TELOMERES IN HUMAN HEALTH

While it is clear that telomeres play an important role in human health, there is significant variability in the extent to which aberrant telomere biology is connected to the disease. For example, patients with the prototypic telomere biology disorder DC are at high risk of bone marrow failure, pulmonary and liver fibrosis, immunodeficiency, and certain cancers.[58,59] Patients with DC have TL below the first percentile for age.

In contrast, association studies of TL and disease in the general population evaluate TL that are significantly different between cases and controls but still within the normal range of TL (i.e., not as short as DC). As discussed above, studies of the general population suggest that short telomeres are associated with mortality,[14,60] cancer,[15] cardiovascular diseases,[16] infections ([17]), and inflammatory disorders.[61] But similar to all epidemiology research, the question remains "do shorter than average telomeres directly cause health problems in the general population?"

Making a valid causal inference from observational epidemiological studies is complicated mainly by possible bias associated with study design or the presence of uncontrolled (or unknown) confounding. A confounding factor in an exposure–outcome relationship is a variable that is associated with both the exposure and the outcome.[62] Age is an example of a common confounder in observational studies of age-related diseases such as cancer, where the observed association is affected by the unbalanced age distribution between comparison groups. Figure 5.2 provides a schematic description of the relationship between TL (exposure), age (confounder), and cancer (outcome). In population studies, age only explains a small percentage of the observed interindividual variation in TL, ranging from 8%[63] to 29%,[64] but it is still an important confounder in TL studies. For instance, in a recent study evaluating the role of TL in cancer risk in a large population-based study, age adjustment contributed to a 70% attenuation

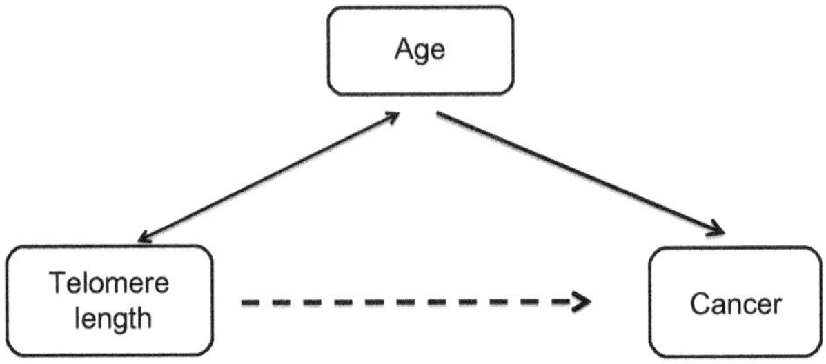

Figure 5.2 A schematic diagram describing confounding. The bidirectional arrow represents a noncausal association, the unidirectional arrow represents a causal association, and the dashed line represents the observed association.

in the calculated hazard ratio (unadjusted hazard ratio; HR = 1.74 vs. adjusted HR = 0.98) of cancer incidence in individuals with short versus long telomeres.[63] TL is also associated with sex,[65] race,[4] BMI,[66] and tobacco smoking,[12] many of which are known risk factor for age-related diseases and therefore have to be accounted for when evaluating the relationship between TL and such diseases.

While confounding may mislead a true association and should be eliminated, valid interacting factors can be of biological importance. Interaction or effect modification describes a relationship between two or more factors, in which they alter the effect of each other with respect to the outcome.[62] Some important examples of interacting factors have been noted in TL epidemiology studies. Fu and colleagues found that liver cirrhosis modified the association between hepatocellular carcinoma (HCC) and leukocyte TL.[67] Liver cirrhosis is a strong risk factor for HCC in patients with chronic hepatitis C and B viruses. In that study, significant associations between leukocyte telomeres and HCC risk were restricted to noncirrhotic patients (OR = 3.54, 95% CI 1.58–7.93 vs. OR = 0.95, 95% CI 0.55–1.64, in patients with liver cirrhosis). Identifying interactions with biological relevance can provide a risk stratification tool that could be valuable for screening or treatment strategies. In another example, Puterman *et al.*[68] found that physical activity can modify the observed inverse association between TL and perceived stress, suggesting that exercise can protect human cells from the negative biological effect of stress, as reflected by TL. A subsequent intervention trial suggested that lifestyle changes, including stress management and physical activity, may affect TL over time.[19] Testing for interactions is important, but, as in all studies, the biological plausibility of the interactions and the findings needs to be considered.

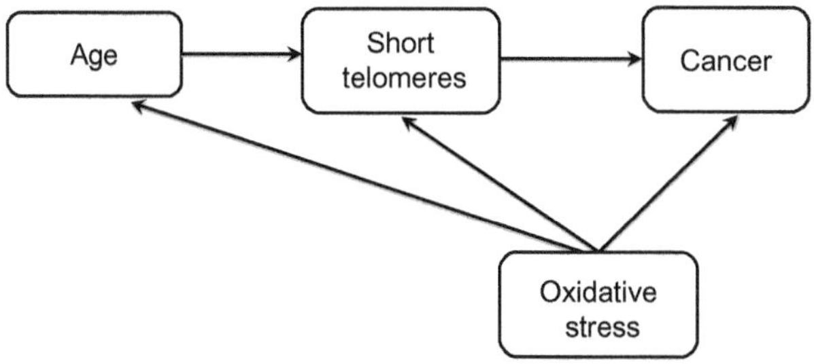

Figure 5.3 A schematic diagram describing mediation where short telomeres are in the causal pathway of age and cancer.

Evaluating the relationship between TL, risk factors, and diseases in a multidimensional way is attractive for an in-depth understanding of the actual role TL plays in human diseases. One example is through mediation analysis.[69] The mediation model identifies factors that are in the causal pathway between the study variable and outcome. Through it, a researcher can quantify the total effect of an exposure on the outcome of interest and can calculate how much of this association is through the mediator of interest. Figure 5.3 illustrates a theoretical example in which TL is in the causal pathway between age and cancer and extends this to show that this model can accommodate multiple factors (e.g., oxidative stress). Several investigators have suggested that telomere dysfunction may causally mediate the association between psychological stress and cardiovascular diseases[70] or lifestyle factors and risk of type 2 diabetes.[71] Mediation analysis has been used in few small studies to evaluate if TL could explain some of the epidemiological observations that are not yet well understood including the relationship between childhood socioeconomic status and the high risk of respiratory infections in adulthood[72] and a possible connection between consanguinity and mental illnesses.[73]

5. THE EFFECT OF TL METHOD OF MEASUREMENT ON STUDY RESULTS

Several methods have been used to measure TL, each with its strengths and limitations reviewed in detail elsewhere.[74–76] In this section, we focus on issues that may impact validity of TL–disease association studies comparing the three most commonly used TL measurement methods in epidemiological and clinical research. Southern blots of terminal restriction fragments (TRF) are considered the gold standard for TL measurement and are used to calibrate other methods.[77] TL has also been successfully measured using real-time quantitative polymerase chain reaction (qPCR).[78,79] Both Southern blot and qPCR techniques measure TL in extracted DNA and, therefore, can be used for stored samples commonly available for epidemiological studies. However, qPCR is more attractive for large-scale studies because it is high-throughput, needs small amounts of DNA, and is relatively inexpensive. Both methods are limited by their reported lack of sensitivity to detect very short telomeres and by the fact that their TL measurement is an average across population of cells per sample.[76,80] On the other hand, flow cytometry with fluorescent *in situ* hybridization (flow FISH) TL measurement of leukocytes is very accurate and provides cell population-specific measure,[81] but its utility in population-based studies is limited by its need for viable cells.

The correlation of TL values between these three measurement methods is generally modest. In an epidemiological study of 681 elderly individuals, the correlation coefficient (r) between TL data generated by Southern blot and qPCR was weak ($r=0.52$).[41] This is significantly weaker than reported correlations between TL measured by the two methods in the initial validation studies of either qPCR monoplex assay ($r=0.82$)[78] or qPCR monochrome multiplex assay ($r=0.91$).[79] Our data comparing TL measured by qPCR and flow FISH in 52 normal individuals showed a similar result ($r=0.47$).[82] No data are available for the correlation between flow FISH and Southern blot TL beyond the strong correlation reported in the initial flow FISH study ($r=0.9$).[81]

The weak correlations between methods may partially explain some of the inconsistencies observed in the published literature when different TL measurement methods are used. For example, a large cohort study measuring relative TL using qPCR reported significant excess in cancer mortality associated with short telomeres ($n=787$, HR$=11.1$ in the shortest vs. longest tertile).[83] However, a second study[84] measuring TL using Southern blot analysis reported no association between TL and cancer mortality ($n=1{,}136$; HR$=1.2$, in the shortest vs. longest quartile). In a meta-analysis including TL–obesity studies, the authors reported that the results from studies that used the Southern blot assay to measure TL were more consistent than those that used qPCR.[66] Similar inconsistencies between studies evaluating the association between qPCR-relative TL and breast cancer have been observed (two showed an association with long telomeres, three showed an association with short telomeres, and three showed no association).[15] Of interest, this variation existed across different study designs (case-control and prospective studies),[15] reflecting a possible role for measurement variability between laboratories and highlighting the need to standardize qPCR laboratory procedure in relative TL measurement. In a study comparing blinded duplicate blood samples for Southern blot TL and qPCR-relative TL, results from Southern blot assay were found to be more consistent than that of qPCR (coefficient of variation (CV) for Southern blot = 1.7% vs. 6.5% for qPCR).[64] Of note, qPCR-relative TL measurements may be sensitive to DNA quality, quantity, and extraction method.[85]

These differences in TL measurement methods illustrate the importance of minimizing technique-related measurement error and thus limit misclassification bias. Simple measures that can help with this include (1) distributing samples from both comparisons groups (case/control and exposed/unexposed) on the same plate; (2) avoiding the use of DNA samples of unknown source or unknown extraction method; (3) and using a single method of DNA extraction or, at least, matching the cases and controls on the extraction method.

Of note, recent research suggests a need to switch from measuring average TL to a more biologically relevant measurement of TL based on the proportion of short telomeres in cells.[76] *In vitro* and animal studies suggest that this measurement may be a better indicator of cell aging than the average TL, which can be effected by the presence of very long telomeres in some cells.[86,87]

6. ANALYTIC CONSIDERATIONS IN TL EPIDEMIOLOGY STUDIES

Determining the number of individuals included in a study is one of the important early decisions made during the study design. Things to be taken into account when calculating sample size include[88] study design, scale of measurement of the response variables (i.e., are we comparing means or proportions?), the difference we want to detect (larger sample sizes are needed to detect smaller differences), the population interindividual variability of the measure of interest (the more homogeneous the population, the smaller the number of individuals needed), the desired level of statistical significance (the smaller the p-value, the larger the sample size needed), and study power (i.e., probability to detect differences when they exist; the higher the desired power, the larger the sample size needed).

For TL studies, a large enough sample size is especially important because interindividual TL heterogeneity and age variability must be included in the calculations. Using simulations, it was shown that longitudinal designs may have more power to observe TL association with age-related diseases, because they can capture the actual TL attrition rate.[43]

Adjusting for confounders is crucial in epidemiological studies. Several methods are used—some are at the design level and others are used during the analysis. At the design level, comparison groups (e.g., cases and controls) are often matched by potential confounders (e.g., age). A disadvantage of a matched design is that strict matching (one-to-one) limits our ability to evaluate the relationship of the matched variable with the study outcome. At the analysis level, regression models are the most commonly used method to control for confounding. However, it is important to note that limitations may exist (reviewed in detail in Ref. 89) including the following: (1) The adjusted effect size can be invalid if the distribution of the potential confounder greatly differs between studied groups (e.g., when certain age groups are not presented in the control groups, usually very young or very old); (2) multivariable model misspecification can lead to biased results, such as when the model doesn't account for some important nonlinearities. Nonlinearity has been observed in TL relationship with age in several studies.[39,64] For instance, a study of 835 individuals found that a piecewise linear curve with breakpoints hinged at 1 and 18 years of age provided the best fit to the data as compared with linear, quadratic, or cubic curves.[39]

Another widely used method to control for confounding, mostly to adjust for age, is calculating age-standardized TL using an independent population.[5,7] Again, selecting a large enough reference population for each age category is important to ensure capturing the TL variability.

7. CONCLUSIONS

Epidemiological studies are crucial in understanding the role of telomere biology in human aging and disease. In this chapter, we have summarized some important epidemiological concepts and issues that can affect the interpretation of the results from epidemiological studies of TL. Careful selection of an appropriate study design, an adequate sample size, and an accurate TL measurement assay are important factors that can significantly affect the findings of the study. Future research to compare TL measurement from different assays in large studies is needed. This will aid the development of more reliable assays for large epidemiological studies. It is also paramount to measure TL from multiple samples collected in longitudinal epidemiological studies with long follow-up times to precisely estimate the age-related telomere attrition rate and how it relates to disease risk. Including large enough sample from different age groups is necessary to enhance our understanding of telomere dynamics during the life span. Telomere measurement in somatic tissues will provide information on the organ-specific aging rate. Additionally, standardization of available measurement methods is crucial to allow comparing between studies. Recent research calls for new methods that measure the proportion of short telomeres within cells rather than the current average TL measurement. Such an assay could provide many research opportunities and help to better understand the role of telomeres in health and disease.

In summary, proper epidemiological research has the potential to clarify the role telomeres play in disease risk and may eventually guide preventive strategies, identify modifiable targets, and aid in the development of diagnostic and prognostic biomarkers.

ACKNOWLEDGMENT

This work is supported by the Intramural Research Program of the Division of Cancer Epidemiology and Genetics, National Cancer Institute, National Institutes of Health.

REFERENCES

1. Harley CB, Futcher AB, Greider CW. Telomeres shorten during ageing of human fibroblasts. *Nature*. 1990;345:458–460.
2. Warters M, Griffen AB. The telomeres of Drosophila. *J Hered*. 1950;41:182–190.
3. Frenck Jr RW, Blackburn EH, Shannon KM. The rate of telomere sequence loss in human leukocytes varies with age. *Proc Natl Acad Sci U S A*. 1998;95:5607–5610.
4. Hunt SC, Chen W, Gardner JP, et al. Leukocyte telomeres are longer in African Americans than in whites: the National Heart, Lung, and Blood Institute Family Heart Study and the Bogalusa Heart Study. *Aging Cell*. 2008;7:451–458.
5. Alter BP, Baerlocher GM, Savage SA, et al. Very short telomere length by flow fluorescence in situ hybridization identifies patients with dyskeratosis congenita. *Blood*. 2007;110:1439–1447.
6. Calado RT, Brudno J, Mehta P, et al. Constitutional telomerase mutations are genetic risk factors for cirrhosis. *Hepatology*. 2011;53:1600–1607.
7. Alter BP, Rosenberg PS, Giri N, Baerlocher GM, Lansdorp PM, Savage SA. Telomere length is associated with disease severity and declines with age in dyskeratosis congenita. *Haematologica*. 2012;97:353–359.
8. Drury SS, Theall K, Gleason MM, et al. Telomere length and early severe social deprivation: linking early adversity and cellular aging. *Mol Psychiatry*. 2012;17(7):719–727.
9. Epel ES, Blackburn EH, Lin J, et al. Accelerated telomere shortening in response to life stress. *Proc Natl Acad Sci U S A*. 2004;101:17312–17315.
10. Krauss J, Farzaneh-Far R, Puterman E, et al. Physical fitness and telomere length in patients with coronary heart disease: findings from the Heart and Soul Study. *PLoS One*. 2011;6:e26983.

11. Mirabello L, Huang WY, Wong JY, et al. The association between leukocyte telomere length and cigarette smoking, dietary and physical variables, and risk of prostate cancer. *Aging Cell.* 2009;8:405–413.
12. Valdes AM, Andrew T, Gardner JP, et al. Obesity, cigarette smoking, and telomere length in women. *Lancet.* 2005;366:662–664.
13. Zhang X, Lin S, Funk WE, Hou L. Environmental and occupational exposure to chemicals and telomere length in human studies. *Postgrad Med J.* 2013;89:722–728.
14. Cawthon RM, Smith KR, O'Brien E, Sivatchenko A, Kerber RA. Association between telomere length in blood and mortality in people aged 60 years or older. *Lancet.* 2003;361:393–395.
15. Wentzensen IM, Mirabello L, Pfeiffer RM, Savage SA. The association of telomere length and cancer: a meta-analysis. *Cancer Epidemiol Biomarkers Prev.* 2011;20:1238–1250.
16. Fitzpatrick AL, Kronmal RA, Gardner JP, et al. Leukocyte telomere length and cardiovascular disease in the cardiovascular health study. *Am J Epidemiol.* 2007;165:14–21.
17. Ilmonen P, Kotrschal A, Penn DJ. Telomere attrition due to infection. *PLoS One.* 2008;3:e2143.
18. Jacobs TL, Epel ES, Lin J, et al. Intensive meditation training, immune cell telomerase activity, and psychological mediators. *Psychoneuroendocrinology.* 2011;36:664–681.
19. Ornish D, Lin J, Chan JM, et al. Effect of comprehensive lifestyle changes on telomerase activity and telomere length in men with biopsy-proven low-risk prostate cancer: 5-year follow-up of a descriptive pilot study. *Lancet Oncol.* 2013;14:1112–1120.
20. Rothman K. *Epidemiology an Introduction.* New York, NY: Oxford University Press; 2002.
21. Porta M. *A Dictionary of Epidemiology.* 5th ed. New York, NY: Oxford University Press; 2008.
22. Aubert G, Lansdorp PM. Telomeres and aging. *Physiol Rev.* 2008;88:557–579.
23. Daniali L, Benetos A, Susser E, et al. Telomeres shorten at equivalent rates in somatic tissues of adults. *Nat Commun.* 2013;4:1597.
24. Friedrich U, Griese E-U, Schwab M, Fritz P, Thon K-P, Klotz U. Telomere length in different tissues of elderly patients. *Mech Ageing Dev.* 2000;119:89–99.
25. Takubo K, Izumiyama-Shimomura N, Honma N, et al. Telomere lengths are characteristic in each human individual. *Exp Gerontol.* 2002;37:523–531.
26. Thomas P, O'Callaghan NJ, Fenech M. Telomere length in white blood cells, buccal cells and brain tissue and its variation with ageing and Alzheimer's disease. *Mech Ageing Dev.* 2008;129:183–190.
27. Gadalla SM, Cawthon R, Giri N, Alter BP, Savage SA. Telomere length in blood, buccal cells, and fibroblasts from patients with inherited bone marrow failure syndromes. *Aging (Albany NY).* 2010;2:867–874.
28. Heaphy CM, Yoon GS, Peskoe SB, et al. Prostate cancer cell telomere length variability and stromal cell telomere length as prognostic markers for metastasis and death. *Cancer Discov.* 2013;3:1130–1141.
29. Meeker AK, Hicks JL, Gabrielson E, Strauss WM, De Marzo AM, Argani P. Telomere shortening occurs in subsets of normal breast epithelium as well as in situ and invasive carcinoma. *Am J Pathol.* 2004;164:925–935.
30. Heaphy CM, Subhawong AP, Gross AL, et al. Shorter telomeres in luminal B, HER-2 and triple-negative breast cancer subtypes. *Mod Pathol.* 2011;24:194–200.
31. Heaphy CM, Bisoffi M, Fordyce CA, et al. Telomere DNA content and allelic imbalance demonstrate field cancerization in histologically normal tissue adjacent to breast tumors. *Int J Cancer.* 2006;119:108–116.
32. Kurabayashi R, Takubo K, Aida J, et al. Luminal and cancer cells in the breast show more rapid telomere shortening than myoepithelial cells and fibroblasts. *Hum Pathol.* 2008;39:1647–1655.

33. Zhou X, Meeker AK, Makambi KH, et al. Telomere length variation in normal epithelial cells adjacent to tumor: potential biomarker for breast cancer local recurrence. *Carcinogenesis*. 2012;33:113–118.
34. O'Sullivan JN, Bronner MP, Brentnall TA, et al. Chromosomal instability in ulcerative colitis is related to telomere shortening. *Nat Genet*. 2002;32:280–284.
35. Rampazzo E, Bertorelle R, Serra L, et al. Relationship between telomere shortening, genetic instability, and site of tumour origin in colorectal cancers. *Br J Cancer*. 2010;102:1300–1305.
36. O'Sullivan J, Risques RA, Mandelson MT, et al. Telomere length in the colon declines with age: a relation to colorectal cancer? *Cancer Epidemiol Biomarkers Prev*. 2006;15:573–577.
37. Broer L, Codd V, Nyholt DR, et al. Meta-analysis of telomere length in 19,713 subjects reveals high heritability, stronger maternal inheritance and a paternal age effect. *Eur J Hum Genet*. 2013;21:1163–1168.
38. Sanders JL, Newman AB. Telomere length in epidemiology: a biomarker of aging, age-related disease, both, or neither? *Epidemiol Rev*. 2013;35(1):112–131.
39. Aubert G, Baerlocher GM, Vulto I, Poon SS, Lansdorp PM. Collapse of telomere homeostasis in hematopoietic cells caused by heterozygous mutations in telomerase genes. *PLoS Genet*. 2012;8:e1002696.
40. Moller P, Mayer S, Mattfeldt T, Muller K, Wiegand P, Bruderlein S. Sex-related differences in length and erosion dynamics of human telomeres favor females. *Aging (Albany NY)*. 2009;1:733–739.
41. Elbers CC, Garcia ME, Kimura M, et al. Comparison between southern blots and qPCR analysis of leukocyte telomere length in the health ABC study. *J Gerontol A Biol Sci Med Sci*. 2014;69(5):527–531.
42. Cassidy A, De Vivo I, Liu Y, et al. Associations between diet, lifestyle factors, and telomere length in women. *Am J Clin Nutr*. 2010;91:1273–1280.
43. Aviv A, Valdes AM, Spector TD. Human telomere biology: pitfalls of moving from the laboratory to epidemiology. *Int J Epidemiol*. 2006;35:1424–1429.
44. Zhang W, Chen Y, Yang X, et al. Functional haplotypes of the hTERT gene, leukocyte telomere length shortening, and the risk of peripheral arterial disease. *PLoS One*. 2012;7:e47029.
45. Bojesen SE, Pooley KA, Johnatty SE, et al. Multiple independent variants at the TERT locus are associated with telomere length and risks of breast and ovarian cancer. *Nat Genet*. 2013;45:371–384.
46. Kote-Jarai Z, Saunders EJ, Leongamornlert DA, et al. Fine-mapping identifies multiple prostate cancer risk loci at 5p15, one of which associates with TERT expression. *Hum Mol Genet*. 2013;22:2520–2528.
47. Burnett-Hartman AN, Fitzpatrick AL, Kronmal RA, et al. Telomere-associated polymorphisms correlate with cardiovascular disease mortality in Caucasian women: the Cardiovascular Health Study. *Mech Ageing Dev*. 2012;133:275–281.
48. Yan L, Wu S, Zhang S, Ji G, Gu A. Genetic variants in telomerase reverse transcriptase (TERT) and telomerase-associated protein 1 (TEP1) and the risk of male infertility. *Gene*. 2014;534:139–143.
49. Hartwig FP. Telomere length and telomere-related genetic variations in epidemiology: getting the context right. *J Genet Syndr Gene Ther*. 2013;4:150.
50. Pooley KA, Bojesen SE, Weischer M, et al. A genome-wide association scan (GWAS) for mean telomere length within the COGS project: identified loci show little association with hormone-related cancer risk. *Hum Mol Genet*. 2013;22:5056–5064.
51. Codd V, Nelson CP, Albrecht E, et al. Identification of seven loci affecting mean telomere length and their association with disease. *Nat Genet*. 2013;45:422–442, 7e1–2.

52. Rothman KJ. Types of epidemiologic study. In: Rothman K, ed. *Epidemiology: An Introduction*. New York, NY: Oxford University Press; 2002:78–84.
53. Hou L, Zhang X, Gawron AJ, Liu J. Surrogate tissue telomere length and cancer risk: shorter or longer? *Cancer Lett.* 2012;319:130–135.
54. Aviv A, Chen W, Gardner JP, et al. Leukocyte telomere dynamics: longitudinal findings among young adults in the Bogalusa Heart Study. *Am J Epidemiol.* 2009;169:323–329.
55. Chen W, Kimura M, Kim S, et al. Longitudinal versus cross-sectional evaluations of leukocyte telomere length dynamics: age-dependent telomere shortening is the rule. *J Gerontol A Biol Sci Med Sci.* 2011;66A:312–319.
56. Chen W, Kimura M, Kim S, et al. Longitudinal versus cross-sectional evaluations of leukocyte telomere length dynamics: age-dependent telomere shortening is the rule. *J Gerontol A Biol Sci Med Sci.* 2011;66:312–319.
57. Steenstrup T, Hjelmborg JV, Kark JD, Christensen K, Aviv A. The telomere lengthening conundrum—artifact or biology? *Nucleic Acids Res.* 2013;41:e131.
58. Vulliamy T, Dokal I. Dyskeratosis congenita. *Semin Hematol.* 2006;43:157–166.
59. Ballew BJ, Savage SA. Updates on the biology and management of dyskeratosis congenita and related telomere biology disorders. *Expert Rev Hematol.* 2013;6:327–337.
60. Zhong R, Liu L, Zou L, et al. Genetic variations in TERT-CLPTM1L locus are associated with risk of lung cancer in Chinese population. *Mol Carcinog.* 2013;52(Suppl 1): E118–E126.
61. Kinouchi Y, Hiwatashi N, Chida M, et al. Telomere shortening in the colonic mucosa of patients with ulcerative colitis. *J Gastroenterol.* 1998;33:343–348.
62. Gordis L. *Epidemiology*. 2nd ed. Pennsylvania: W.B. Saunders Company; 2000.
63. Weischer M, Nordestgaard BG, Cawthon RM, Freiberg JJ, Tybjaerg-Hansen A, Bojesen SE. Short telomere length, cancer survival, and cancer risk in 47102 individuals. *J Natl Cancer Inst.* 2013;105:459–468.
64. Aviv A, Hunt SC, Lin J, Cao X, Kimura M, Blackburn E. Impartial comparative analysis of measurement of leukocyte telomere length/DNA content by Southern blots and qPCR. *Nucleic Acids Res.* 2011;39:e134.
65. Gardner M, Bann D, Wiley L, et al. Gender and telomere length: systematic review and meta-analysis. *Exp Gerontol.* 2013;51C:15–27.
66. Muezzinler A, Zaineddin AK, Brenner H. Body mass index and leukocyte telomere length in adults: a systematic review and meta-analysis. *Obes Rev.* 2014;15(3):192–201.
67. Fu X, Wan S, Hann HW, et al. Relative telomere length: a novel non-invasive biomarker for the risk of non-cirrhotic hepatocellular carcinoma in patients with chronic hepatitis B infection. *Eur J Cancer.* 2012;48:1014–1022.
68. Puterman E, Lin J, Blackburn E, O'Donovan A, Adler N, Epel E. The power of exercise: buffering the effect of chronic stress on telomere length. *PLoS One.* 2010;5:e10837.
69. Richiardi L, Bellocco R, Zugna D. Mediation analysis in epidemiology: methods, interpretation and bias. *Int J Epidemiol.* 2013;42(5):1511–1519.
70. Epel ES, Lin J, Wilhelm FH, et al. Cell aging in relation to stress arousal and cardiovascular disease risk factors. *Psychoneuroendocrinology.* 2006;31:277–287.
71. Elks CE, Scott RA. The long and short of telomere length and diabetes. *Diabetes.* 2014;63:65–67.
72. Cohen S, Janicki-Deverts D, Turner RB, et al. Childhood socioeconomic status, telomere length, and susceptibility to upper respiratory infection. *Brain Behav Immun.* 2013;34:31–38.
73. Mansour H, Chowdari K, Fathi W, et al. Does telomere length mediate associations between inbreeding and increased risk for bipolar I disorder and schizophrenia? *Psychiatry Res.* 2011;188:129–132.
74. Aubert G, Hills M, Lansdorp PM. Telomere length measurement-caveats and a critical assessment of the available technologies and tools. *Mutat Res.* 2012;730:59–67.

75. Lin KW, Yan J. The telomere length dynamic and methods of its assessment. *J Cell Mol Med*. 2005;9:977–989.
76. Vera E, Blasco MA. Beyond average: potential for measurement of short telomeres. *Aging (Albany NY)*. 2012;4:379–392.
77. Aviv A. Commentary: raising the bar on telomere epidemiology. *Int J Epidemiol*. 2009;38:1735–1736.
78. Cawthon RM. Telomere measurement by quantitative PCR. *Nucleic Acids Res*. 2002; 30:e47.
79. Cawthon RM. Telomere length measurement by a novel monochrome multiplex quantitative PCR method. *Nucleic Acids Res*. 2009;37:e21.
80. Horn T, Robertson BC, Gemmell NJ. The use of telomere length in ecology and evolutionary biology. *Heredity (Edinb)*. 2010;105:497–506.
81. Rufer N, Dragowska W, Thornbury G, Roosnek E, Lansdorp PM. Telomere length dynamics in human lymphocyte subpopulations measured by flow cytometry. *Nat Biotechnol*. 1998;16:743–747.
82. Khincha PP, Gadalla SM, Katki H, et al. Evaluating the utility of telomere length measurement by qPCR as a diagnostic test for dyskeratosis congenita. *Blood*. 2013;122:3700.
83. Willeit P, Willeit J, Mayr A, et al. Telomere length and risk of incident cancer and cancer mortality. *JAMA*. 2010;304:69–75.
84. Fitzpatrick AL, Kronmal RA, Kimura M, et al. Leukocyte telomere length and mortality in the Cardiovascular Health Study. *J Gerontol A Biol Sci Med Sci*. 2011;66:421–429.
85. Cunningham JM, Johnson RA, Litzelman K, et al. Telomere length varies by DNA extraction method: implications for epidemiologic research. *Cancer Epidemiol Biomarkers Prev*. 2013;22:2047–2054.
86. Berardinelli F, Nieri D, Sgura A, Tanzarella C, Antoccia A. Telomere loss, not average telomere length, confers radiosensitivity to TK6-irradiated cells. *Mutat Res*. 2012;740:13–20.
87. Vera E, Bernardes de Jesus B, Foronda M, Flores JM, Blasco MA. The rate of increase of short telomeres predicts longevity in mammals. *Cell Rep*. 2012;2:732–737.
88. Rigby AS, Vail A. Statistical methods in epidemiology. II: a commonsense approach to sample size estimation. *Disabil Rehabil*. 1998;20:405–410.
89. Martin W. Making valid causal inferences from observational data. *Prev Vet Med*. 2014;113(3):281–297.

CHAPTER SIX

Telomere Dysfunction and Hematologic Disorders

Raquel M.A. Paiva, Rodrigo T. Calado
Department of Internal Medicine, University of São Paulo at Ribeirão Preto School of Medicine, Ribeirão Preto, São Paulo, Brazil

Contents

Abstract

Aplastic anemia is a disease in which the hematopoietic stem cell fails to adequately produce peripheral blood cells, causing pancytopenia. In some cases of acquired aplastic anemia and in inherited type of aplastic anemia, dyskeratosis congenita, telomere biology gene mutations and telomere shortening are etiologic. Telomere erosion hampers the ability of hematopoietic stem and progenitor cells to adequately replicate, clinically resulting in bone marrow failure. Additionally, telomerase mutations and short telomeres are genetic risk factors for the development of some hematologic cancers, including myelodysplastic syndrome, acute myeloid leukemia, and chronic lymphocytic leukemia.

1. INTRODUCTION

Defective telomere maintenance provides the molecular mechanism for some human bone marrow failure syndromes. Telomeres were first

Progress in Molecular Biology and Translational Science, Volume 125
ISSN 1877-1173
http://dx.doi.org/10.1016/B978-0-12-397898-1.00006-2

described to be short in peripheral blood leukocytes of patients with aplastic anemia (AA).[1] In this first and pivotal report, telomere erosion in peripheral blood leukocytes worsened in patients with persistent cytopenias and inversely correlated with time of disease, whereas telomeres were stable in patients who achieved complete remission after treatment, suggesting telomere length as a marker for the hematopoietic stem cell (HSC) reserve. Telomere attrition appeared to be a surrogate for the excessive cell divisions to which HSCs were submitted, the "mitotic clock." However, at that time, it was not clear whether telomere erosion was merely a consequence of HSC exhaustion, due to the destruction of the bone marrow, or rather played a more important part in the etiology of HSC failure and AA. This question was answered when mutations in the *DKC1* gene, which encodes dyskerin, were identified in dyskeratosis congenita (DC), an inherited type of AA.[2] Based on studies in *DKC1* orthologs, dyskerin function was then thought to be related to ribosome biogenesis. It was only later that dyskerin was found to be a component of the telomerase complex,[3] thus implicating telomere shortening as etiologic in bone marrow failure syndromes rather than a consequence (Fig. 6.1).

DC is caused by defects in telomere maintenance genes leading to excessive telomere erosion, but telomere defects also permeates the pathogenesis of a wider variety of bone marrow failure syndromes. Mutations in

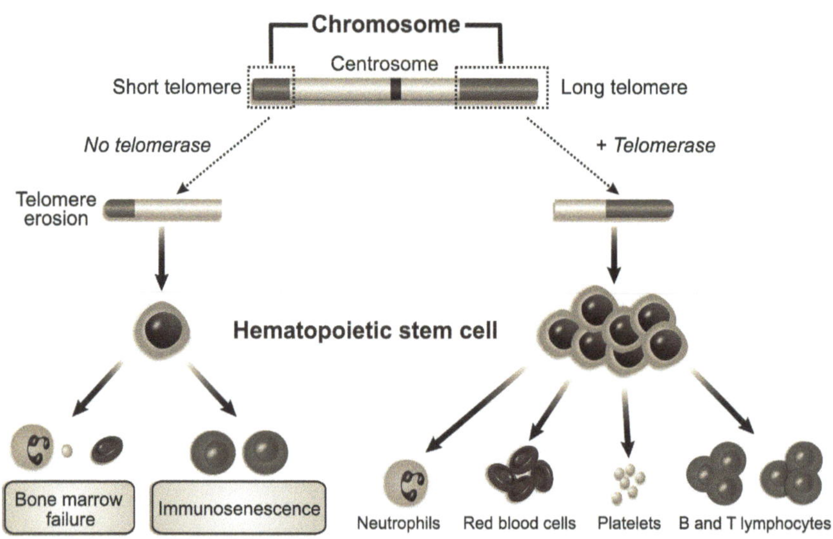

Figure 6.1 Short telomeres engage hematopoietic stem cells into senescence and apoptosis, impairing the appropriate production of peripheral blood cells.

telomerase and telomere biology components also may be identified in some patients with apparently acquired AA, idiopathic pulmonary fibrosis, and liver diseases not presenting the classical manifestations of DC.[4–7] Defective telomere maintenance is a new category of diseases sharing a common molecular mechanism, generally called "telomere diseases" or "telomeropathies."[8]

Telomere diseases include a large spectrum of clinical presentations in both children and adults of which bone marrow failure is the most recognized manifestation. Additionally, the severity of disease varies, from profound and early marrow failure associated with other physical abnormalities, as observed in DC, Hoyeraal–Hreidarsson and Revesz syndromes, to less severe phenotypes restricted to the hematopoietic compartment, as seen in AA. The degree of telomere attrition also appears to correlate with disease severity, as the shortest telomeres are observed in the former syndromes. Red cell, granulocyte, and platelet counts in peripheral blood may be affected by telomere erosion and patients with short telomeres are less likely to respond to immunosuppressive therapies (ISTs).[9,10] Telomere shortening also may be observed in other congenital AAs, such as Fanconi anemia (FA)[11,12] and Shwachman–Diamond syndrome (SDS)[13,14]; however, the contribution of telomere dysfunction to disease development in these instances is less direct. Additionally, telomere shortening and telomerase mutations also are risk factors for hematologic cancers, such as myelodysplastic syndrome (MDS), acute myeloid leukemia (AML), and chronic lymphocytic leukemia (CLL).

In general, a wide spectrum of phenotypes may be observed in patients with short telomeres. Telomere shortening may participate as modulator of disease severity and in the onset of symptoms and signs earlier or later in life. Thus, telomere-mediated diseases are directly associated to short telomere length, genetic, and epigenetic modifiers.[15,16] Etiologic factors, such as bone marrow stress, aging, and mutations in telomerase complex genes, may induce telomere erosion with pathological consequences for humans. Excessive telomere shortening results in defective cell proliferation, senescence, apoptosis, genomic instability, breakage–fusion–bridge cycles, and aneuploidy and the clinical consequences are a direct result of these molecular events. Environmental factors may also contribute to telomere shortening as well as injury to an organ with limited regeneration capacity, thus potentially triggering telomere-mediated disease.[17]

This chapter reviews the contribution of dysfunctional telomeres to hematologic disorders.

2. BONE MARROW FAILURE

AA is a group of inherited and acquired hematologic disorders characterized by the quantitative and qualitative impairment of HSCs in the bone marrow, resulting in a hypoplastic marrow and low peripheral blood cell counts (pancytopenia).[18] In AA, HSCs are not capable of producing the daily demand for red cells, leukocytes, and platelets. In acquired AA, an immune-mediated destruction of hematopoietic stem and progenitor cells is the major mechanism of disease and it affects both children and adults. Clinical manifestations are due to low blood cell counts: anemia, infection, and mucocutaneous bleeding. Inherited forms of AA usually manifest in the first decade of life and a variety of gene defects compromising HSC function are etiologic. In inherited AA, in addition to hematologic manifestations, physical anomalies also may be observed, such as growth retardation, organ impairment, skin lesions, endocrine insufficiency, and a proclivity for cancer development. FA is the most common type of constitutional AA in which lesions in genes involved in DNA repair (FA pathway) cause marrow failure. DC is the second most common type of inherited AA in which abnormal telomere repair is etiologic. SDS is another type of constitutional AA in which marrow failure associates with exocrine pancreatic deficiency.

Abnormal telomere maintenance plays a role in disease development in both acquired and inherited forms of AA. DC is a more severe phenotype with higher penetrance and the prototype for telomere diseases.

3. DYSKERATOSIS CONGENITA

DC is the second most common type of inherited bone marrow failure syndromes (IBMFS) and is named after its clinical manifestation of ectodermal dysplasia, a triad of dystrophic nails, patchy skin hyperpigmentation, and oral leukoplakia.[19] DC was first described by Zinsser[20] in early 1900s and was recognized as a clinical entity by Engman[21] and Cole[22]. Although the mucocutaneous changes of ectodermal dysplasia are the most common findings in DC, bone marrow failure is found in approximately 80% of patients with DC and is the main cause of death.[19,23] Patients with DC also present other clinical features, such as an increased risk for cancer development, in particular squamous cell carcinoma of the tongue, AML, MDS, and impairment in other organs, mainly hepatic cirrhosis and pulmonary fibrosis, which are additional life-threatening manifestations of DC.[8,23–25]

Progressive pulmonary fibrosis develops in around 10–15% of patients with DC.[26,27] Other clinical manifestations are esophageal stricture, epiphora, short stature, and growth retardation.

Bone marrow failure manifests in the first decades of life, and hematopoietic stem cell transplantation (HSCT) is the only curative therapy for marrow failure, in which the defective HSCs are replaced with allogeneic HSCs.[28,29] However, the success of HSCT is restricted to hematopoiesis and it does not decrease the risk of cancer development. Additionally, transplant may trigger the disease in other organs, mainly the liver and the lungs, as posttransplant complications affecting these organs are more common in DC than in other hematologic disorders.[30] Not all patients with DC have a suitable donor for transplant and other therapeutic options to alleviate marrow failure are necessary. Treatment with androgens may induce hematologic response in up to 60% of patients with DC.[19] Male sex hormones directly stimulate erythropoiesis in humans and increase the red cell mass. Additionally, sex hormones also activate telomerase in hematopoietic cells[31] but stimulating the *TERT* gene expression, as estrogen receptor elements are present in its promoter region. Male hormones appear to act on *TERT* expression via its conversion into estrogens by aromatase. However, the effects of androgens on other affected organs or in cancer susceptibility are not known.

Three different inheritance patterns have been described in DC: X-linked, autosomal dominant (AD), and autosomal recessive. In large registries, the most common inheritance pattern reported is X-linked, affecting the *DKC1* gene, which encodes dyskerin, a component of the telomerase complex. The A353V mutation is the most common in humans causing X-linked DC.[32,33] It is found in 40% of cases and is often found as a *de novo* mutation.[2,17,23,32] Dyskerin is a nucleolar protein (58 kDa) associated with small nucleolar RNAs (snoRNAs) in H/ACA small nucleolar ribonucleoprotein complexes.[23,34,35] Telomerase RNA (TERC) contains an H/ACA snoRNA-like sequence at its 3′ end that binds dyskerin and three other H/ACA proteins. Box H/ACA, characteristic of mammalian TERC, may assemble telomerase in Cajal bodies and its translocation to the telomeres.[3,36,37] Indeed, dyskerin physically associates with telomerase RNA, which is an integral part of telomerase enzyme acting as template for the synthesis of the TTAGGG repeats found at the telomeres of all linear chromosomes.[3,38] This physical interaction may explain telomere erosion and reduced telomerase activity in patient's leukocytes.[39] Autosomal dominant dyskeratosis, caused by mutations or large deletions in *TERC*, is present in about 5–10% of DC families,[23,32,33,40] and low frequency of *TERC* mutations may be found in patients with severe aplastic anemia (SAA) and MDS who had received a bone marrow transplant.[41] Inherited AD DC is also found with *TERT* mutations, caused by single amino acid substitutions characterizing severe disease and anticipation with shorter telomeres in later generations.[23,29] Some cases of DC are linked to autosomal recessive mutations. Mutations in *TCAB1*, also known as *WDR79*, *WRAP53*, revealed defective telomerase trafficking as a cause of autosomal recessive DC.[42] *TCAB1* is responsible to mediate the localization of specific H/ACA class RNAs, including hTR, to Cajal bodies through recognition of a CAB box, crucial phenomenon for telomere length maintenance.[29,37,43] *TCAB1* heterozygous mutations may be found in patients with classical DC presenting the mucocutaneous triad, bone marrow failure, and telomere lengths below the first percentile compared to those found in healthy controls.[23,29] NOP10 and NHP2 have also been implicated in autosomal recessive DC. Both are associated with dyskerin and with RNA telomerase pertaining of the active telomerase complex.[44,45] Individuals with NOP10 or NHP2 mutations may present classical DC features, with severe telomere shortening and reduced RNA telomerase levels.[29]

In contrast to the mentioned genes, mutations in *TINF2*, *CTC1*, and *RTEL1* genes do not appear to impact telomere length through direct effects on telomerase.[46–49] Sporadic (*de novo*) AD DC is found in 15% of all probands with mutations in *TINF2*, which encodes TIN2, a component of shelterin complex, presenting high penetrance with *R282H* and *R282C* most common mutations. *TINF2* mutations may be characterized in severe to classical DC with dramatic telomere shortening and early age of onset disease and are more likely to present with bone marrow failure prior to manifesting any sign of the classic triad. *TINF2* mutations also have been identified in some patients with severe DC variants Revesz and Hoyeraal–Hreidarsson syndromes and, thus far, are the only gene mutations associated with Revesz syndrome.[17,23,29,46]

In addition to shelterin, a trimeric complex (CST) consisting of conserved telomere maintenance component 1 (CTC1), suppressor of cdc13-1 (STN1), and telomeric pathway with STN1 (TEN1) has been proposed to contribute to telomere maintenance and homeostasis.[49–51] The CST complex has been implicated in telomerase activity inhibition *in vitro*, and CST knockdown may result in telomerase-dependent telomere elongation *in vivo*.[49,52] CTC1 also may promote the restart of telomere lagging strand synthesis at stalled replication forks through interaction with DNA polymerase α primase.[49,53,54]

Mutations in CTC1 were identified in patients with a rare, autosomal recessive disorder Coats plus, a form of cerebroretinal microangiopathy with calcifications and cysts that includes extraneurologic manifestations, including osteopenia and gastrointestinal bleeding.[47,48] Although mucocutaneous features and bone marrow failure, characteristic of DC, are less typical of the Coats plus disorder, CTC1 mutations were identified in several patients with classic DC.[49]

Novel mutations were identified in *RTEL1* gene in patients with severe variant of DC. *RTEL1* is a DNA helicase, essential for genome stability that has been implicated as an antirecombinase in that it disrupts D-loop formation during homologous recombination[55–58] and is essential for the disassembly of T-loops during DNA replication.[59] *RTEL1* is crucial for DNA repair and telomere maintenance.[57,58] Mutations in *RTEL1* were identified in a small cohort of patients with severe features of DC, including patients with Hoyeraal–Hreidarsson syndrome (HHS) and classical DC. HHS is a phenotypically severe variant of DC. It has been characterized as multisystem bone marrow failure syndrome with defective telomere maintenance. Biallelic mutations of *RTEL1* were found in 6/23 index cases with HHS, and telomere lengths were significantly shorter in cases than in controls. This group presented significantly higher levels of telomeric circles, produced as a consequence of incorrect processing of telomere ends, than did controls. Cell harboring *RTEL1* mutations have significant defects in telomere maintenance, but not in homologous recombination, and incorrect resolution of T-loops may be a mechanism for telomere shortening and telomere-mediated disease in humans.[58]

In some rare instances, DC may be independent of telomere shortening. Twenty DC families, clinically classified as DC, were recently identified with homozygous mutations in *C16orf57* gene.[60] Individuals with *C16orf57* mutations present multisystem abnormalities including abnormal skin pigmentation/poikiloderma at a young age, neutropenia, nail dystrophy, and pulmonary disease.[60] Differently of the previously characterized cases of DC with mutations in *DKC1*, *TERC*, *TERT*, *NOP10*, *NHP2*, *TCAB1*, and *TINF2* and association with short telomeres, patients with *C16orf57* mutations, presenting features of DC, do not present short telomere length, suggesting that telomere length maintenance is unaffected in this group of patients.[60] Additional reports with DC-variant HHS contribute to the notion that some cases of DC or its variants may be due to telomere length-independent telomere dysfunction. In a particular study, the patient expressed a unique splice variant of Apollo, a 5′ to 3′ exonuclease implicated in both intrastrand cross-link repair and telomere protection.[61] Patient-derived primary dermal fibroblasts exhibited increased telomere dysfunction-induced foci, despite normal bulk telomere length. Similarly, intriguing results have been reported in another patient with HHS.[62] The primary dermal fibroblasts from this patient had increased telomere dysfunction-induced foci yet normal telomere length; however, leukocyte telomere length was measured and found to be severely short in the patient and affected siblings. One proposed explanation for the differences in telomere lengths is that it relates to differences in the replicative demand on dermal fibroblasts versus peripheral blood leukocytes. The gene encoding Apollo was not sequenced in this patient, although the genes are known to be mutated in DC were found to be normal as was the telomerase activity level.[29,62] Further studies are necessary to comprise the DC disease with telomere-independent mechanism.

4. DISEASE ANTICIPATION

Patients with DC presenting defects in the telomerase complex with expected defective telomere maintenance and short telomeres have been shown with disease anticipation (increasing severity and/or earlier onset) in succeeding generations.[63] Anticipation may be explained by simultaneous inheritance of the mutation and of chromosomes with short telomeres.[17,63]

Disease anticipation has been observed in several generation families with AD DC.[33,64–66] Large kindreds with AD DC, presenting mutations in *TERC*, may exhibit disease phenotype at younger ages in successive generations, and siblings that did not inherit a mutated TERC copy may not have early onset of the clinical symptoms, but they inherit shorter telomeres from the affected parent and may also be at risk.[17,66–68] Furthermore, the penetrance, severity, and time of onset of the clinical features of DC are variable even among family members with the same mutation.[66]

4.1. Abnormal rRNA modification and dyskerin function

Telomere shortening, caused by mutation in genes from telomerase complex, may not be the sole molecular mechanism responsible for DC phenotype.[17]

Cbf5p, dyskerin's yeast ortholog, is essential for pre-rRNA processing and is related to participate in rRNA pseudouridylation suggesting dyskeratosis as a ribosomal disease in early observations. A mutant dyskerin might become rate limiting for pre-rRNA processing and therefore for ribosome biosynthesis, or it could be that inadequately processed rRNA makes ribosomes functionally abnormal.[69]

Alterations in ribosome modification are also related to cancer and other features of the disease. Studies in an animal model hypomorphic *Dkc1^{m}* mice showed defective rRNA pseudouridylation in early generations[70] with abnormal accumulation of immature rRNA species. In another study, *Dkc1^{m}* mice presented defects in IRES (internal ribosome entry site), impairing translation of messenger RNAs containing IRES elements such as p27(Kip1) and the antiapoptotic factors Bcl-xL and X-linked inhibitor of apoptosis protein, indicating that defective ribosome activity may lead to disease and cancer.[71]

In addition, products of genes mutated in other syndromes, such as cartilage-hair hypoplasia, Diamond–Blackfan anemia (DBA), and SDS, are all predicted to be involved in aspects of ribosome synthesis.[72–76] DBA is characterized by a defect of erythroid progenitors and, clinically, by anemia and malformations. About 50% of DBA patients are related to present one of the four most commonly mutated genes: *RPS19*, *RPL5*, *RPL11*, and *RPS26*. The *RPS19* gene is the most frequently mutated, affecting 25% of patients. Mutations in *RPL5* and *RPL11* are frequently found in patients with malformations of upper limbs or face.[77]

Cells of patients with X-linked DC do not show abnormal rRNA processing or rRNA accumulation. Patients with mutations at telomerase complex (*TERC*, *TERT*, or *NOP10*) are not predicted to have abnormal rRNA pseudouridylation, despite clinical features similar to X-linked DC.[17,26] Further studies are regarded to elucidate the evidence of abnormal ribosomal modification in the pathophysiology of DC.[17,78]

5. APLASTIC ANEMIA

AA is diagnosed based on marrow hypoplasia and low peripheral blood counts with at least two cytopenias: hemoglobin <10 g/dL, neutrophils <1500/μL, and/or platelets <50,000/μL.[79] Although the "empty" BM defines AA, the BM is not truly empty, but replaced by fat cells.[80,81] Blood counts can be decreased in infections and diseases as systemic lupus, but severe persistent pancytopenia has a limited differential diagnosis and almost always implies BM involvement. AA may develop at any age and, based on underlying etiology, it is classified into two different forms: inherited and acquired. Acquired AA is characterized by an immune-mediated mechanism, and in a minority of cases, it may be attributable to exposure to environmental risk factors; chemical compounds; idiosyncratic reactions to drugs, such as chloramphenicol; and certain infections that may contribute to the etiology of acquired AA.[66,79] In most patients, the destruction of hematopoiesis in the bone marrow is due to an immune mechanism dominated by activated cytotoxic T lymphocytes. These cells target hematopoietic CD34+ stem cells and drastically diminish their number in the bone marrow. Since the first reports of hematopoiesis recovery in patients who fail engraftment after stem cell transplant, immunosuppression was inferred to treat AA. The responsiveness to ISTs in this disease is the major evidence of an immune-mediated process. Failure to respond to such treatments might be due to a nonimmune etiology or to properties of the hematopoietic compartment, a quantitatively severe loss of stem cell number, or a qualitative abnormality of hematopoietic stem and progenitor cells.[79,82] The reason for T-cell activation in AA patients is still unclear. However, a combination of genetic risk factors, including SNPs in tumor necrosis factor-alpha and gamma-interferon, may contribute to this spectrum.[66,83] The clinical relevance of varying telomere lengths (TL) and/or mutations in genes of the telomerase complex (*TERC* and *TERT*) is evolving in AA.[82] AA has been seen earlier in most patients with DC, but also can be the only manifestation of a spectrum of telomere syndromes. The distinction between acquired and inherited forms has relevance in clinical practice, becoming critical to treatment decisions.[49,84,85]

Initial reports described mutations in the *TERC* gene in patients with apparently acquired AA who lacked the typical physical stigmata of DC and an obvious family history.[17,86,87] Functional consequences as reduced telomerase activity of patient primary cells, very short telomeres of leukocytes, and low telomerase activity *in vitro* were observed in patients with genetic changes in *TERC* gene.[17,88]

Mutations in *TERC* and *TERT* have been identified in 5–10% of acquired SAA cases that present later in life without physical abnormalities.[8,82] Initial studies suggested that up to one-third of patients with SAA have short telomere.[1,10] Short telomeres were initially presumed to be secondary to hematopoietic "stress"; however, the discovery of loss-of-function mutations in genes of the telomerase complex (TERC and TERT) established a genetic etiology for telomere attrition in some patients with marrow failure who did not have the physical anomalies associated with an IBMFS.[82,86,87] Genetic variants and specific haplotypes for genes coding for shelterin components as TRF1 and TRF2 might contribute to marrow failures. Functionally, TRF1 and TRF2 bind directly to the telomere DNA strand from damage as well as regulating telomere length, and mutations reducing their expression or impairing their binding to DNA result in telomere erosion.[17,89] Thus, the role of telomeres in acquired AA has been of considerable interest.[49]

AA can be treated with hematopoietic stem cell transplant (HSCT) or IST. HSCT is curative in the majority of cases, but many patients do not have a compatible donor or are not suitable for transplant. Moreover, IST is the first-line treatment for many patients, especially those over age 40 years. IST with antithymocyte globulin plus cyclosporine partially recovers blood counts in 60–75% of patients in most large studies, suggesting that the hematopoietic progenitor and stem cell depletion is most often immune-mediated.[90] However, not all patients respond to immunosuppression; some lose their response over time, and some experience evolution to MDS or AML. Thus, there is no marker to predict the outcome following HSCT or IST that could reduce risks for patients, allowing a rational treatment allocation.[82] Treatments with androgens may reduce the clonal evolution in AA patients.[91,92] Androgen therapy might be associated with hematologic improvements along with TL (telomere length) elongation of peripheral blood leukocytes.[82,93] In this context, modulation of telomere attrition may reduce clonal events in humans, and the effects of androgen therapy on increasing telomerase activity might contribute to slowing telomere attrition, assisting in cell regeneration and ultimately preventing genomic instability.[82,92,94]

6. CLINICAL MANAGEMENT IN BONE MARROW FAILURE SYNDROMES

AA is the major cause of death (80%) in patients with DC and is a common manifestation in other bone marrow failure diseases.[5] Target therapies for DC are complicated due to the highly variable onset and progression of DC manifestations, even among family members.[66,95,96] Patients may present classical physical findings and normal bone marrow function, and others may present at very young ages with the diagnostic triad and early-onset bone marrow failure.[66] For patients with bone marrow failure in general, hematopoietic stem cell transplantation (HSCT) is the front-line treatment, but complications are frequently reported as pulmonary fibrosis, hepatic, and gastrointestinal diseases.[66,97–99]

The main application of telomere length in AA patients is referred to distinguish acquired (immune-mediated) and inherited forms (mutations in *TERC* or *TERT*). The differential diagnosis is essential for correct therapeutic actions.[100] Both forms of AA present similar clinical characteristics; however, for the acquired form, immunosuppression may be considered as treatment. Inherited AA patients usually do not respond to immunosuppression and, in this case, bone marrow transplantation is the first line of treatment.[49] In the most severe cases of AA, telomere length may be taken into account in the therapeutic decision between immunosuppression and bone marrow transplantation. The decision to get hematopoietic stem cell transplantation or immunosuppression as the first therapy in SAA is currently based on age and availability of a histocompatible donor and the ability to predict hematologic response; relapse and clonal evolution could improve treatment allocation. Telomere length is not a biomarker for late events in SAA; however, it is involved in the destabilization of the genome and participates in the etiopathogenesis of clonal evolution.[82]

Androgen therapy may also be considered for bone marrow failure, due to direct effect of androgens on *TERT* gene expression and telomerase activity. *In vitro* studies with human peripheral blood leukocytes and bone marrow cells exposed to androgens have shown to increase telomerase activity along with higher expression of *TERT* mRNA.[82,92,101] This direct androgen action may explain hormone stimulation of hematopoiesis in DC and in acquired AA patients, leading to hematopoietic response by activating telomerase in hematopoietic cells, specially stem cells.[92] Telomere shortening observed in leukocytes is frequently observed in all the bone marrow failure syndromes, acquired or constitutional. In the setting of bone marrow failure, short telomere length may be a surrogate marker for DC. Flow-fluorescence *in situ* hybridization (flow-FISH) is able to distinguish DC patients from their unaffected relatives and from patients with other marrow failure syndromes. Abnormally short telomeres of lymphocytes showed 91% sensitivity and specificity for the diagnosis of DC, and the analysis of abnormal telomeres of granulocytes was more sensitive (96%) but less specific (84%).[17,102] Another sensitive technique is the single telomere length assay (STELA), which has been used for the identification of critically short telomeres regardless of the average TL by qPCR. STELA assay is based on a single PCR reaction to amplify chromosome-specific telomeres.[103] A recent study observed that very short telomeres, detected by STELA, were present in a higher proportion of SAA patients who progressed to MDS when compared with complete responders to immunosuppression.[104]

Other techniques such as Southern blot, considered the gold standard for TL analysis, qPCR, and FISH also may be employed for telomere length measurement.[105–108]

As very short telomeres may be a risk for chromosomal instability and clonal evolution in some AA patients, new methodologies to detect telomere length have been performed for rapid diagnosis and treatment of diseases with bone marrow failure.

7. TELOMERE SHORTENING IN OTHER BONE MARROW FAILURE SYNDROMES

In addition to DC, two other forms of IBMFS are frequent, namely, Fanconi anemia (FA) and Shwachman–Diamond syndrome (SDS).

SDS is a rare autosomal recessive inherited syndrome characterized by bone marrow failure, pancreatic exocrine insufficiency, skeletal defects, and predisposition to myelodysplasia and leukemia.[14,78,109] More than 90% of Shwachman–Diamond cases are caused by biallelic mutations in *SBDS*, located at chromosome 7.[110–112] *SBDS* gene is suggested to be involved in RNA processing, and the protein's nucleolar localization in mammalian cells further supports this role.[14,113] Mutations in affected patients result from gene conversion between *SBDS* and an *SBDSP* pseudogene, the two most common being insertion of a premature stop codon and splice site disruption with resultant translation truncation.[14,74] In humans, SBDS deficiency does not result in changes in the rRNA maturation profile, although SBDS may coprecipitate with 28S rRNA.[114] SBDS deficiency predisposes to marrow failure and some patients may present shortened telomeres. However, no physical association between the SBDS protein and the telomerase complex is observed[14] and telomere attrition in SDS appears to be an epiphenomenon.

FA is a chromosome-instability syndrome clinically characterized by bone marrow failure, developmental defects, and physical anomalies, with increased risk to cancer.[115] The disease is genetically determined by mutations in 16 different genes (*FANC-A*, *FANC-B*, *FANC-C*, *FANC-D1*, *FANC-D2*, *FANC-E*, *FANC-F*, *FANC-G*, *FANC-I*, *FANC-J*, *FANC-L*, *FANC-M*, *FANC-N*, *FANC-O*, *FANC-P*, and *FANC-Q*), and all FA proteins cooperate in the FA pathway to repair cross-linked DNA.[116] Although the exact mechanism of *FANC* gene remains unclear, most FA genes interact in pathways that recognize and repair DNA damage, and bone marrow failure may be attributed to inadequate DNA repair, triggering of apoptosis and genomic intability.[117] In addition, recent work relates that most *FANC* genes have additional functions in other DNA repair transactions and some of them have been found to modulate cytokine responses or interact with redox-sensing proteins, independently of their role in DNA repair.[116,118,119] Although they have very few similarities, the encoded FA proteins cooperate in a common FA/BRCA pathway by forming several complexes, where the activation of a key FA protein FANCD2 (and FANCI) seems to orchestrate the cascade of events in response to DNA damage.[101,120]

FA patients develop bone marrow failure culminating in pancytopenia. Patients, at the median age of 7 years, start presenting hematologic complications, such as low platelet count, followed by leukopenia, before eventually developing AA.[116,121,122] As all blood lineages become strongly deficient, it implies HSC dysfunction, and FA patients present very low numbers of CD34+ cells.[116,123,124] Recent work has shown significant depletion of the CD34+ cell fraction in very young FA patients, well before the onset of pancytopenia, and this very low number of CD34 cells in young FA patients points to a prenatal origin of the stem cell defect.[123,125,126] The very low number of CD34+ cells in young FA patients points to a prenatal origin of the stem cell defect. In fact, knockdown of FA genes in human embryonic stem cells results in defective hematopoiesis.[116]

Telomeres may be shorter in peripheral blood mononuclear cells of FA patients. Erosion of telomeres is prognostic, directly correlating with the severity of the marrow failure, the probability of developing frank AA, the presence of organ malformation, and, inversely, with the age of onset for myelodysplasia and tumors.[11] Telomere loss may be the result of direct telomeric DNA damage followed by inadequate DNA repair; telomere damage in FA might be excessive due to increased production of reactive oxygen species[127] known to cause telomere erosion.[128] Excess telomere breaks were observed in FA cells, correlating with extra-chromosomal telomeres.[129] Since telomerase activity is preserved in FA patients, FA proteins appear to have a function in ALT (alternative lengthening of telomere) telomere maintenance; FANCD2 colocalizes with telomeres in various ALT cell lines,[130–132] and depletion of FANCA or FANCD2 inhibits homologous recombination required for telomere maintenance in these cells.[130] Because ALT cells have an activated DNA damage response at dysfunctional telomeres,[133] the function of the FA pathway at ALT may be related to its general role in DNA damage responses.[132]

FANCD2 interacts with telomeric DNA and telomeric protein TNKS1. In normal conditions, FANCD2 may be considered as safeguards of genomic integrity by inhibiting TNKS1-mediated TRF1 PARsylation and by protecting telomeric DNA through TRF1121.

The role of *FANCC* was examined in telomere length regulation and telomere recombination *in vivo*.[134] Murine model with depletion of *Fancc* did not directly affect telomere length or endcapping in a strain with long telomere, but an increase of telomere sister chromatid exchange in the late generations of telomerase-mutant mice with short telomeres was observed. These findings support that *FANCC* does not directly regulate long telomeres, but regulates short telomere-initiated telomere recombination and ablation of *FANCC* function and may promote telomere recombination.[134]

8. ANIMAL MODELS

Telomeres of normal laboratory mice (*Mus musculus*) are much longer than in humans, despite a much shorter murine life span.[17] Murine models have been created to "knock out" TERT (*Tert*$^{-/-}$)[135] or the RNA component (*Terc*$^{-/-}$)[136] and downregulate DKC1 (*Dkc1*^{m}).[70] In addition, other investigators have used the CAST/EiJ mouse strain, which has a shorter telomere length.[137] Homozygous mice are viable, but after few generations, as telomeres shorten, they become infertile.[17] *Terc*$^{-/-}$ mice also have splenic atrophy, a reduced lymphocyte proliferative capacity, and impaired hematopoietic function, although overt cytopenia is not observed.[138] Fourth-generation *Terc*$^{-/-}$ murine cells show signs of chromosomal instability[139]; in addition to consistent telomere shortening and the appearance of chromosomes lacking telomere capping, there are aneuploidy and other chromosomal abnormalities, including end-to-end fusions[135] and an increased rate of tumor formation.[138]

Although murine models may not be the ideal organisms to study human model telomere biology, the relationship between telomere shortening and risk of disease has been largely examined in experimental model of mice. Rudolph *et al.*[140] examined telomere shortening and risk of cirrhosis in mice null for telomerase reverse transcriptase. Telomerase-deficient mice, null for the essential telomerase RNA (mTR) gene, were subjected to genetic, surgical, and chemical ablation of the liver, and telomere dysfunction was associated with defects in liver regeneration and accelerated the development of liver cirrhosis in response to chronic exposure of CCl_4.[140] Experimental evidence in murine model (*Terc*$^{-/-}$) has provided the existence of an *in vivo* process of cellular senescence induced by critical telomere shortening with functional impact on organ regeneration.[141] In addition, mice telomere dysfunction (*Tert*$^{-/-}$) has been checked to repress PGC network and mitochondrial biogenesis induced by p53, thereby linking telomeres to mitochondrial biology, oxidative defense, and metabolism.[142] Mice hepatocytes with Dkc1 deletion present signs of liver damage including an increase serum alanine aminotransferase activity and a disordered structure at the histological and macroscopic levels. Also, after CCl_4 administration, hepatocytes from wild-type mice mount a rapid proliferative response, instead those without dyskerin. Therefore, hepatocytes can survive without dyskerin but the role of dyskerin in RNA modification is essential for cellular proliferation.[143] The findings in experimental animals and observations in humans suggest that reduced telomerase activity may contribute to the development of diseases.

9. TELOMERES AND HEMATOLOGIC CANCERS

Telomere attrition is presumed to increase the risk of cancer due to chromosomal instability eventually leading to aneuploidy caused by aberrant chromosomal segregation during cell division, chromosomal translocation, and abnormal nucleotide sequences, all relevant process for cancer formation, progression, or malignant transformation.[17,49,144,145] Damaged and/or short telomeres can be recognized as DNA double-strand breaks and activation of the DNA repair machinery. Telomeres become the targets of nonhomologous end joining or homology-directed repair, facilitating aneuploidy, inversions, translocations, and deletions.[17,31,146,147]

The increased cancer risk and predisposition to MDS and AML in DC and AA patients are presumed to be related to the excessive telomere shortening, due to the absence of a DNA damage response.[49,148]

Those patients with DC have a high risk of developing AML and that *TERT* mutations are observed in patients with hematologic malignancies implicated telomere dysfunction in leukemogenesis.[31,49,149] In a cohort of 133 consecutive AML patients, 8% of patients were found to carry variants in the *TERT* gene when compared to controls.[150] Mutations were constitutional and impaired telomerase enzymatic function. The *TERT* A1062T variant was the most common SNP in AML patients, and later, it also was found at higher frequencies in patients with CLL and large B-cell lymphoma.[49,150,151] Additionally, patients with AA and shorter telomeres are at higher risk for evolving to MDS and AML after treatment.[152] Bone marrow cells of patients with AA and short telomeres display indirect evidence for chromosomal instability: telomere-free ends, aneuploidy, and nonclonal chromosomal abnormalities, such as translocations, deletions, insertions, and chromosome association.[16] Taken together, these findings suggest that telomerase mutations and telomere attrition lead to senescence and apoptosis of HSC clones with short telomeres and at the same time that it positively selects stem cell chromosomal changes and more prone for malignant transformation.

Most recently, CLL patients with short telomeres in malignant cells presented genetic aberrations and evidence of DNA damage at telomeres.[149] In a large cohort of CLL patients, short telomeres and high telomerase profile were independently associated with more rapid disease progression. High telomerase activity and short telomeres were prognostic markers of disease progression and worse clinical outcome. CLL cases with high telomerase levels and short telomeres were frequently characterized by an unmutated immunoglobulin heavy-chain variable (*IGVH*) gene status and high-risk chromosomal aberrations. Conversely, CLL cases with low telomerase levels and long telomeres were associated with mutated *IGVH* and low-risk chromosome abnormalities. The presence or absence of somatic mutations in the IGVH gene is considered a good prognostic discriminator in CLL. Unmutated IGVH profile is associated with an aggressive clinical course.[149] In another study on CLL, recurrent somatic mutations were detected in the gene encoding shelterin complex POT1. Around 3.5% of CLL patients presented *POT1* mutations and *POT1* mutated CLL cells presented numerous telomeric and chromosomal abnormalities when compared to wild-type cells, suggesting *POT*1 mutations as a factor for the malignant features of CLL cells.[153] Telomere fusions also have been observed to occur in CLL as disease progresses, suggesting that telomere crisis plays a role in leukemic evolution.[154]

Short telomeres may be a marker for cancer susceptibility.[155] Pancreatic ductal adenocarcinoma (PDAC) patients are related to extensive telomere dysfunction and chromosomal abnormalities that persist even after metastatic dissemination. Interestingly, telomere shortening is detectable at the earliest stages of pancreatic cancer formation and may precede the acquisition of KRAS oncogene mutations. Mutations in KRAS oncogenes represent around 90% of all PDACs and are considered an early event in the progression from normal ductal cells to invasive adenocarcinoma.[156]

Telomere shortening has emerged as a hallmark of cirrhotic liver and consequent hepatocellular carcinoma regardless of telomerase mutations. Telomere shortening is linked to replicative senescence in hepatocytes of cirrhotic liver with the presence of markers such as p21, p27, and SA-β-galactosidase.[157] Telomerase mutations resulting in telomere erosion are considered genetic risk factor for human cirrhosis and may predispose affected subjects to disease progression in combination with environmental injury.[6]

Thus, constitutional hypomorphic telomerase mutations and short telomeres provide a genetic basis for cancer predisposition. Other nongenetic causes for excessive telomere erosion also might contribute to genomic instability. Etiologic factors as stem cell transplantation or chemotherapy, smoking, psychological stress, and mutations in telomerase complex genes and environmental factors are associated to telomere shortening resulting in defective cell proliferation, senescence, apoptosis, genomic instability, breakage–fusion–bridge cycles, and aneuploidy, triggering telomere-mediated disease development.[17,31,44]

REFERENCES

1. Ball SE, Gibson FM, Rizzo S, Tooze JA, Marsh JC, Gordon-Smith EC. Progressive telomere shortening in aplastic anemia. *Blood.* 1998;91:3582–3592.
2. Heiss NS, Knight SW, Vulliamy TJ, et al. X-linked dyskeratosis congenita is caused by mutations in a highly conserved gene with putative nucleolar functions. *Nat Genet.* 1998;19:32–38.
3. Mitchell JR, Wood E, Collins K. A telomerase component is defective in the human disease dyskeratosis congenital. *Nature.* 1999;402:551–555.
4. Yamaguchi H, Calado RT, Ly H, et al. Mutations in TERT, the gene for telomerase reverse transcriptase, in aplastic anemia. *N Engl J Med.* 2005;352:1413–1424.
5. Armanios M. Syndromes of telomere shortening. *Annu Rev Genomics Hum Genet.* 2009;10:45–61.
6. Calado RT, Brudno J, Mehta P, et al. Constitutional telomerase mutations are genetic risk factors for cirrhosis. *Hepatology.* 2011;53:1600–1607.
7. Hartmann D, Srivastava U, Thaler M, et al. Telomerase gene mutations are associated with cirrhosis formation. *Hepatology.* 2011;53:1608–1617.
8. Calado RT, Young NS. Telomere diseases. *N Engl J Med.* 2009;361:2353–2365.
9. Lee JJ, Kook H, Chung IJ, et al. Telomere length changes in patients with aplastic anaemia. *Br J Haematol.* 2001;112:1025–1030.
10. Brummendorf TH, Maciejewski JP, Young NS, Lansdorp PL. Telomere length in leukocyte sub-populations of patients with aplastic anemia. *Blood.* 2001;97:895–900.
11. Li X, Leteurtre F, Rocha V, et al. Abnormal telomere metabolism in Fanconi's anaemia correlates with genomic instability and the probability of developing severe aplastic anaemia. *Br J Haematol.* 2003;120:836–845.
12. Leteurtre F, Li X, Guardiola P, et al. Accelerated telomerase shortening and telomerase activation in Fanconi's anemia. *Bone Marrow Transplant.* 1999;23:S30.

13. Thornley I, Dror Y, Sung L, Wynn RF, Freedman MH. Abnormal telomere shortening in leucocytes of children with Shwachman-Diamond syndrome. *Br J Haematol.* 2002;117:189–192.
14. Calado RT, Graf SA, Wilkerson KL, et al. Mutations in the SBDS gene in acquired aplastic anemia. *Blood.* 2007;110:1141–1146.
15. Armanios M, Blackburn EH. The telomere syndromes. *Nat Rev Genet.* 2012;13:693–704.
16. Calado RT, Cooper JN, Padilla-Nash HM, et al. Short telomeres result in chromosomal instability in hematopoietic cells and precede malignant evolution in human aplastic anemia. *Leukemia.* 2012;26:700–707.
17. Calado RT, Young NS. Telomere maintenance and human bone marrow failure. *Blood.* 2008;111:4446–4455.
18. Young NS. Pathophysiologic mechanisms in acquired aplastic anemia. *Hematology Am Soc Hematol Educ Program.* 2006;1:72–77.
19. Dokal I. Dyskeratosis congenita. A disease of premature ageing. *Lancet.* 2001;358:S27.
20. Zinsser F. Atrophia cutis reticularis pigmentione, dystrophia unguium et leukoplakia oris (poikilodermia atrophicans vascularis Jacobi). *Ikonogr Derm.* 1906; 5:219–223.
21. Engman MA. A unique case of reticular pigmentation of the skin with atrophy. *Arch Dermatol Syph.* 1926;13:685–687.
22. Cole H, Rauschkolb J, Toomey J. Dyskeratosis congenita with pigmentation, dystrophia unguis and leukokeratosis. *Arch Dermatol Syph.* 1930;21:71–95.
23. Mason PJ, Bessler M. The genetics of dyskeratosis congenital. *Cancer Genet.* 2011;204:635–645.
24. Knight S, Vulliamy T, Copplestone A, Gluckman E, Mason P, Dokal I. Dyskeratosis congenita (DC) registry: identification of new features of DC. *Br J Haematol.* 1998;103:990–996.
25. Vulliamy T, Dokal I. Dyskeratosis congenita. *Semin Hematol.* 2006;43:157–166.
26. Dokal I. Dyskeratosis congenita in all its forms. *Br J Haematol.* 2000;110:768–779.
27. Walne AJ, Dokal I. Advances in the understanding of dyskeratosis congenita. *Br J Haematol.* 2009;145:164–172.
28. Dietz AC, Orchard PJ, Baker KS, et al. Disease-specific hematopoietic cell transplantation: nonmyeloablative conditioning regimen for dyskeratosis congenita. *Bone Marrow Transplant.* 2011;46:98–104.
29. Nelson ND, Bertuch AA. Dyskeratosis congenita as a disorder of telomere maintenance. *Mutat Res.* 2012;730:43–51.
30. Rocha V, Devergie A, Socié G, et al. Unusual complications after bone marrow transplantation for dyskeratosis congenita. *Br J Haematol.* 1998;103:243–248.
31. Calado RT, Regal JA, Kleiner DE, et al. A spectrum of severe familial liver disorders associate with telomerase mutations. *PLoS One.* 2009;4:7926.
32. Knight SW, Heiss NS, Vulliamy TJ, et al. X-linked dyskeratosis congenita is predominantly caused by missense mutations in the DKC1 gene. *Am J Hum Genet.* 1999;65:50–58.
33. Vulliamy TJ, Marrone A, Knight SW, Walne A, Mason PJ, Dokal I. Mutations in dyskeratosis congenita: their impact on telomere length and the diversity of clinical presentation. *Blood.* 2006;107:2680–2685.
34. Meier UT, Blobel G. NAP57, a mammalian nucleolar protein with a putative homolog in yeast and bacteria. *J Cell Biol.* 1994;127:1505–1514.
35. Meier UT. The many facets of H/ACA ribonucleoproteins. *Chromosoma.* 2005;114:1–14.
36. Fu D, Collins K. Human telomerase and Cajal body ribonucleoproteins share a unique specificity of Sm protein association. *Genes Dev.* 2006;20:531–536.

37. Venteicher AS, Abreu EB, Meng Z, et al. A human telomerase holoenzyme protein required for Cajal body localization and telomere synthesis. *Science*. 2009;323:644–648.
38. Zhao Z, Pan X, Liu L, Liu N. Telomere length maintenance, shortening, and lengthening. *J Cell Physiol*. 2013; http://dx.doi.org/10.1002/jcp.24537.
39. Vulliamy TJ, Knight SW, Mason PJ, Dokal I. Very short telomeres in the peripheral blood of patients with X-linked and autosomal dyskeratosis congenita. *Blood Cells Mol Dis*. 2001;27:353–357.
40. Dokal I, Vulliamy T, Mason P, Bessler M. Clinical utility gene card for: dyskeratosis congenital. *Eur J Hum Genet*. 2011;19. http://dx.doi.org/10.1038/ejhg.2011.90.
41. Field JJ, Mason PJ, An P, et al. Low frequency of telomerase RNA mutations among children with aplastic anemia or myelodysplastic syndrome. *J Pediatr Hematol Oncol*. 2006;28:450–453.
42. Zhong F, Savage SA, Shkreli M, et al. Disruption of telomerase trafficking by TCAB1 mutation causes dyskeratosis congenita. *Genes Dev*. 2011;25:11–16.
43. Tycowski KT, Shu MD, Kukoyi A, Steitz JA. A conserved WD40 protein binds the Cajal body localization signal of scaRNP particles. *Mol Cell*. 2009;34:47–57.
44. Cohen SB, Graham ME, Lovrecz GO, Bache N, Robinson PJ, Reddel RR. Protein composition of catalytically active human telomerase from immortal cells. *Science*. 2007;315:1850–1853.
45. Kiss T, Fayet-Lebaron E, Jady BE. Box H/ACA small ribonucleoproteins. *Mol Cell*. 2010;37:597–606.
46. Savage SA, Alter BP, Chanock SJ, Lansdorp PM. *TINF2*, a component of the shelterin telomere protection complex, is mutated in dyskeratosis congenita. *Am J Hum Genet*. 2008;82:501–509.
47. Anderson BH, Kasher PR, Mayer J, et al. Mutations in CTC1, encoding conserved telomere maintenance component 1, cause Coats plus. *Nat Genet*. 2012;44:338–342.
48. Polvi A, Linnankivi T, Kivela T, et al. Mutations in CTC1, encoding the CTS telomere maintenance complex component 1, cause cerebroretinal microangiopathy with calcifications and cysts. *Am J Hum Genet*. 2012;90:540–549.
49. Gramatges MM, Bertuch AA. Short telomeres: from dyskeratosis congenita to sporadic aplastic anemia and malignancy. *Transl Res*. 2013;162:353–363.
50. Wright WE, Piatyszek MA, Rainey WE, Byrd W, Shay JW. Telomerase activity in human germline and embryonic tissues and cells. *Dev Genet*. 1996;18:173–179.
51. Harrington L. Haploinsufficiency and telomere length homeostasis. *Mutat Res*. 2012;730:37–42.
52. Chen LY, Redon S, Lingner J. The human CST complex is a terminator of telomerase activity. *Nature*. 2012;488:540–544.
53. Stewart JA, Wang F, Chaiken MF, et al. Human CST promotes telomere duplex replication and general replication restart after fork stalling. *EMBO J*. 2012;31:3537–3549.
54. Wang F, Stewart JA, Kasbek C, Zhao Y, Wright WE, Price CM. Human CST has independent functions during telomere duplex replication and C-strand fill-in. *Cell Rep*. 2012;2:1096–1103.
55. Uringa EJ, Youds JL, Lisaingo K, Lansdorp PM, Boulton SJ. RTEL1: an essential helicase for telomere maintenance and the regulation of homologous recombination. *Nucleic Acids Res*. 2011;39:1647–1655.
56. Le Guen T, Jullien L, Schertzer M, et al. RTEL1 (regulator of telomere elongation helicase 1), a DNA helicase essential for genome stability. *Med Sci*. 2013; 29:1138–1144.
57. Lee JW. Telomere shortening by mutations in the RTEL1 helicase cause severe form of dyskeratosis congenita, Hoyerall-Hreidarsson syndrome. *Clin Genet*. 2013;84:210.
58. Walne AJ, Vulliamy T, Kirwan M, Plagnol V, Dokal I. Constitutional mutations in RTEL1 cause severe dyskeratosis congenita. *Am J Hum Genet*. 2013;92:448–453.

59. Vannier JB, Pavicic-Kaltenbrunner V, Petalcorin MI, Ding H, Boulton SJ. RTEL1 dismantles T loops and counteracts telomeric G4-DNA to maintain telomere integrity. *Cell.* 2012;149:795–806.
60. Walne AJ, Vulliamy T, Beswick R, Kirwan M, Dokal I. Mutation in C16orf57 and normal-length telomeres unify a subset of patients with dyskeratosis congenital, poikiloderma with neutropenia and Rothmund-Thomson syndrome. *Hum Mol Genet.* 2010;19:4453–4461.
61. Touzot F, Callebaut I, Soulier J, et al. Function of Apollo (SNM1B) at telomere highlighted by a splice variant identified in a patient with Hoyeraal–Hreidarsson syndrome. *Proc Natl Acad Sci USA.* 2010;107:10097–10102.
62. Lamm N, Ordan E, Shponkin R, Richler C, Aker M, Tzfati Y. Diminished telomeric 3′ overhangs are associated with telomere dysfunction in Hoyeraal–Hreidarsson syndrome. *PLoS One.* 2009;4:5666.
63. Vulliamy T, Marrone A, Szydlo R, Walne A, Mason PJ, Dokal I. Disease anticipation is associated with progressive telomere shortening in families with dyskeratosis congenita due to mutations in TERC. *Nat Genet.* 2004;36(5):447–449.
64. Armanios M, Chen JL, Chang YP, et al. Haploinsufficiency of telomerase reverse transcriptase leads to anticipation in autosomal dominant dyskeratosis congenita. *Proc Natl Acad Sci USA.* 2005;102:15960–15964.
65. Goldman F, Bouarich R, Kulkarni S, et al. The effect of TERC haploinsufficiency on the inheritance of telomere length. *Proc Natl Acad Sci USA.* 2005;102:17119–17124.
66. Savage SA, Bertuch AA. The genetics and clinical manifestations of telomere biology disorders. *Genet Med.* 2010;12:753–764.
67. Marrone A, Walne A, Dokal I. Dyskeratosis congenita: telomerase, telomeres and anticipation. *Curr Opin Genet Dev.* 2005;15:249–257.
68. Vulliamy TJ, Dokal I. Dyskeratosis congenita: the diverse clinical presentation of mutations in the telomerase complex. *Biochimie.* 2008;90:122–130.
69. Luzzatto L, Karadimitris A. Dyskeratosis and ribosomal rebellion. *Nat Genet.* 1998;19:6–7.
70. Ruggero D, Grisendi S, Piazza F, et al. Dyskeratosis congenita and cancer in mice deficient in ribosomal RNA modification. *Science.* 2003;299:259–262.
71. Yoon A, Peng G, Brandenburger Y, et al. Impaired control of IRES-mediated translation in X-linked dyskeratosis congenita. *Science.* 2006;312:902–906.
72. Ridanpää M, van Eenennaam H, Pelin K, et al. Mutations in the RNA component of RNase MRP cause a pleiotropic human disease, cartilage-hair hypoplasia. *Cell.* 2001;104:195–203.
73. Draptchinskaia N, Gustavsson P, Andersson B, et al. The gene encoding ribosomal protein S19 is mutated in Diamond-Blackfan anaemia. *Nat Genet.* 1999;21:169–175.
74. Boocock GR, Morrison JA, Popovic M, et al. Mutations in SBDS are associated with Shwachman–Diamond syndrome. *Nat Genet.* 2003;33:97–101.
75. Myers KC, Davies SM, Shimamura A. Clinical and molecular pathophysiology of Shwachman-Diamond syndrome: an update. *Hematol Oncol Clin North Am.* 2013;27:117–128.
76. Garçon L, Ge J, Manjunath SH, et al. Ribosomal and hematopoietic defects in induced pluripotent stem cells derived from Diamond Blackfan anemia patients. *Blood.* 2013;122:912–921.
77. Boria I, Garelli E, Gazda HT, et al. The ribosomal basis of Diamond-Blackfan anemia: mutation and database update. *Hum Mutat.* 2010;31:1269–1279.
78. Shimamura A. Shwachman-Diamond syndrome. *Semin Hematol.* 2006;43:178–188.
79. Young NS, Scheinberg P, Calado RT. Aplastic anemia. *Curr Opin Hematol.* 2008;15:162–168.

80. Takaku T, Malide D, Chen J, Calado RT, Kajigaya S, Young NS. Hematopoiesis in 3 dimensions: human and murine bone marrow architecture visualized by confocal microscopy. *Blood.* 2010;116:41–55.
81. Young NS. Mouse medicine and human biology. *Semin Hematol.* 2013;50:88–90.
82. Scheinberg P. Prognostic value of telomere attrition in patients with aplastic anemia. *Int J Hematol.* 2013;97:553–557.
83. Demeter J, Messer G, Schrezenmeier H. Clinical relevance of the TNF-alpha promoter/enhancer polymorphism in patients with aplastic anemia. *Ann Hematol.* 2002;81:566–569.
84. Maciejewski JP, Anderson S, Katevas P, Young NS. Phenotypic and functional analysis of bone marrow progenitor cell compartment in bone marrow failure. *Br J Haematol.* 1994;87:227–234.
85. Maciejewski JP, Selleri C, Sato T, Anderson S, Young NS. A severe and consistent deficit in marrow and circulating primitive hematopoietic cells (long-term culture-initiating cells) in acquired aplastic anemia. *Blood.* 1996;88:1983–1991.
86. Fogarty PF, Yamaguchi H, Wiestner A, et al. Late presentation of dyskeratosis congenita as apparently acquired aplastic anaemia due to mutations in telomerase RNA. *Lancet.* 2003;362:1628–1630.
87. Yamaguchi H, Baerlocher GM, Lansdorp PM, et al. Mutations of the human telomerase RNA gene (*TERC*) in aplastic anemia and myelodysplastic syndrome. *Blood.* 2003;102:916–918.
88. Ly H, Calado RT, Allard P, et al. Functional characterization of telomerase RNA variants found in patients with hematological disorders. *Blood.* 2005;105:2332–2339.
89. Savage SA, Calado RT, Xin ZT, Ly H, Young NS, Chanock SJ. Genetic variation in telomeric repeat binding factors 1 and 2 in aplastic anemia. *Exp Hematol.* 2006;34:664–671.
90. Young NS, Calado RT, Scheinberg P. Current concepts in the pathophysiology and treatment of aplastic anemia. *Blood.* 2006;108:2509–2519.
91. Gardner FH, Juneja HS. Androstane therapy to treat aplastic anaemia in adults: an uncontrolled pilot study. *Br J Haematol.* 1987;65:295–300.
92. Calado RT, Yewdell WT, Wilkerson KL, et al. Sex hormones, acting on the TERT gene, increase telomerase activity in human primary hematopoietic cells. *Blood.* 2009;114:2236–2243.
93. Ziegler P, Schrezenmeier H, Akkad J, et al. Telomere elongation and clinical response to androgen treatment in a patient with aplastic anemia and a heterozygous hTERT gene mutation. *Ann Hematol.* 2012;91:1115–1120.
94. Scheinberg P. Aplastic anemia: therapeutic updates in immunosuppression and transplantation. *Hematology Am Soc Hematol Educ Program.* 2012;2012:292–300.
95. Savage SA, Alter BP. Dyskeratosis congenita. *Hematol Oncol Clin North Am.* 2009;23:215–231.
96. Savage SA, Dokal I, Armanios M, et al. Dyskeratosis congenita: the first NIH clinical research workshop. *Pediatr Blood Cancer.* 2009;53:520–523.
97. Dror Y, Freedman MH, Leaker M, et al. Low-intensity hematopoietic stem-cell transplantation across human leucocyte antigen barriers in dyskeratosis congenita. *Bone Marrow Transplant.* 2003;31:847–850.
98. Brazzola P, Duval M, Fournet JC, et al. Fatal diffuse capillaritis after hematopoietic stem-cell transplantation for dyskeratosis congenita despite low-intensity conditioning regimen. *Bone Marrow Transplant.* 2005;36:1103–1105.
99. de la Fuente J, Dokal I. Dyskeratosis congenita: advances in the understanding of the telomerase defect and the role of stem cell transplantation. *Pediatr Transplant.* 2007;11:584–594.

100. Gadalla SM, Savage SA. Telomere biology in hematopoiesis and stem cell transplantation. *Blood Rev.* 2011;25:261–269.
101. Scheinberg P, Young NS. How I treat acquired aplastic anemia. *Blood.* 2012;120:1185–1196.
102. Alter BP, Baerlocher GM, Savage SA, et al. Very short telomere length by flow fluorescence in situ hybridization identifies patients with dyskeratosis congenita. *Blood.* 2007;110:1439–1447.
103. Baird DM. New developments in telomere length analysis. *Exp Gerontol.* 2005;40:363–368.
104. Dumitriu B, Ueda Y, Kajigaya S, Townsley DM, Young NS. Very short telomeres of peripheral blood leukocytes precede clinical progression to myelodysplasia with monosomy 7 in aplastic anemia patients. *ASH Annu Meet Abstr.* 2012;120:1265.
105. Harley CB, Futcher AB, Greider CW. Telomeres shorten during ageing of human fibroblasts. *Nature.* 1990;345:458–460.
106. Cawthon RM. Telomere measurement by quantitative PCR. *Nucleic Acids Res.* 2002;30:47.
107. Lansdorp PM, Verwoerd NP, van de Rijke FM, et al. Heterogeneity in telomere length of human chromosomes. *Hum Mol Genet.* 1996;5:685–691.
108. Poon SS, Martens UM, Ward RK, Lansdorp PM. Telomere length measurements using digital fluorescence microscopy. *Cytometry.* 1999;36:267–278.
109. Dror Y. Shwachman-Diamond syndrome. *Pediatr Blood Cancer.* 2005;45:892–901.
110. Burroughs L, Woolfrey A, Shimamura A. Shwachman-Diamond syndrome: a review of the clinical presentation, molecular pathogenesis, diagnosis, and treatment. *Hematol Oncol Clin North Am.* 2009;23:223–248.
111. Donadieu J, Fenneteau O, Beaupain B, et al. Associated investigators of the French Severe Chronic Neutropenia Registry. Classification of and risk factors for hematologic complications in a French national cohort of 102 patients with Shwachman-Diamond syndrome. *Haematologica.* 2012;97:1312–1319.
112. Calado RT, Dumitriu B. Telomere dynamics in mice and humans. *Semin Hematol.* 2013;50:165–174.
113. Savchenko A, Krogan N, Cort JR, et al. The Shwachman-Bodian-Diamond syndrome protein family is involved in RNA metabolism. *J Biol Chem.* 2005;280:19213–19220.
114. Ganapathi KA, Austin KM, Lee CS, et al. The human Shwachman-Diamond syndrome protein, SBDS, associates with ribosomal RNA. *Blood.* 2007; 110:1458–1465.
115. Bagby G, Alter B. Fanconi anemia. *Semin Hematol.* 2006;43:147–156.
116. Garaycoechea JI, Patel KJ. Why does the bone marrow fail in Fanconi anemia? *Blood.* 2014;123:26–34.
117. Taniguchi T, D'Andrea AD. Molecular pathogenesis of Fanconi anemia: recent progress. *Blood.* 2006;107:4223–4233.
118. Dao KH, Rotelli MD, Petersen CL, et al. FANCL ubiquitinates b-catenin and enhances its nuclear function. *Blood.* 2012;120:323–334.
119. Du W, Adam Z, Rani R, Zhang X, Pang Q. Oxidative stress in Fanconi anemia hematopoiesis and disease progression. *Antioxid Redox Signal.* 2008;10:1909–1921.
120. Lyakhovich A, Ramirez MJ, Castellanos A, et al. Fanconi anemia protein FANCD2 inhibits TRF1 polyADP-ribosylation through tankyrase1-dependent manner. *Genome Integr.* 2011;2:4.
121. Butturini A, Gale RP, Verlander PC, Adler-Brecher B, Gillio AP, Auerbach AD. Hematologic abnormalities in Fanconi anemia: an International Fanconi Anemia Registry study. *Blood.* 1994;84:1650–1655.
122. Kutler DI, Singh B, Satagopan J, et al. A 20-year perspective on the International Fanconi Anemia Registry (IFAR). *Blood.* 2003;101:1249–1256.

123. Lobitz S, Velleuer E. Guido Fanconi (1892–1979): a jack of all trades. *Nat Rev Cancer.* 2006;6:893–898.
124. Kelly PF, Radtke S, von Kalle C, et al. Stem cell collection and gene transfer in Fanconi anemia. *Mol Ther.* 2007;15:211–219.
125. Sejas DP, Rani R, Qiu Y, et al. Inflammatory reactive oxygen species-mediated hematopoietic suppression in Fancc-deficient mice. *J Immunol.* 2007;178:5277–5287.
126. Ceccaldi R, Parmar K, Mouly E, et al. Bone marrow failure in Fanconi anemia is triggered by an exacerbated p53/p21 DNA damage response that impairs hematopoietic stem and progenitor cells. *Cell Stem Cell.* 2012;11:36–49.
127. Korkina LG, Samochatova EV, Maschan AA, et al. Release of active oxygen radicals by leukocytes of Fanconi anemia patients. *J Leukoc Biol.* 1992;52:357–362.
128. Passos JF, Saretzki G, Ahmed S, et al. Mitochondrial dysfunction accounts for the stochastic heterogeneity in telomere-dependent senescence. *PLoS Biol.* 2007;5:110.
129. Callén E, Samper E, Ramírez MJ, et al. Breaks at telomeres and TRF2-independent end fusions in Fanconi anemia. *Hum Mol Genet.* 2002;11:439–444.
130. Fan Q, Zhang F, Barrett B, Ren K, Andreassen PR. A role for monoubiquitinated FANCD2 at telomeres in ALT cells. *Nucleic Acids Res.* 2009;37:1740–1754.
131. Spardy N, Duensing A, Hoskins EE, Wells SI, Duensing S. HPV-16 E7 reveals a link between DNA replication stress, fanconi anemia D2 protein, and alternative lengthening of telomere-associated promyelocytic leukemia bodies. *Cancer Res.* 2008;68:9954–9963.
132. Pang Q, Andreassen PR. Fanconi anemia proteins and endogenous stresses. *Mutat Res.* 2009;668:42–53.
133. Nabetani A, Yokoyama O, Ishikawa F. Localization of hRad9, hHus1, hRad1, and hRad17 and caffeine-sensitive DNA replication at the alternative lengthening of telomeres-associated promyelocytic leukemia body. *J Biol Chem.* 2004;279:25849–25857.
134. Rhee DB, Wang Y, Mizesko M, Zhou F, Haneline L, Liu Y. FANCC suppresses short telomere-initiated telomere sister chromatid exchange. *Hum Mol Genet.* 2010;19:879–887.
135. Blasco MA, Lee HW, Hande MP, et al. Telomere shortening and tumor formation by mouse cells lacking telomerase RNA. *Cell.* 1997;91:25–34.
136. Chiang YJ, Hemann MT, Hathcock KS, et al. Expression of telomerase RNA template, but not telomerase reverse transcriptase, is limiting for telomere length maintenance in vivo. *Mol Cell Biol.* 2004;24:7024–7031.
137. Hao LY, Armanios M, Strong MA, et al. Short telomeres, even in the presence of telomerase, limit tissue renewal capacity. *Cell.* 2005;123:1121–1131.
138. Rudolph KL, Chang S, Lee HW, et al. Longevity, stress response, and cancer in aging telomerase-deficient mice. *Cell.* 1999;96:701–712.
139. Artandi SE, DePinho RA. Mice without telomerase: what can they teach us about human cancer? *Nat Med.* 2000;6:852–855.
140. Rudolph KL, Chang S, Millard M, Schreiber-Agus N, DePinho RA. Inhibition of experimental liver cirrhosis in mice by telomerase gene delivery. *Science.* 2000;287:1253–1258.
141. Satyanarayana A, Wiemann SU, Buer J, et al. Telomere shortening impairs organ regeneration by inhibiting cell cycle re-entry of a subpopulation of cells. *EMBO J.* 2003;22:4003–4013.
142. Sahin E, Colla S, Liesa M, et al. Telomere dysfunction induces metabolic and mitochondrial compromise. *Nature.* 2011;470:359–365.
143. Ge J, Rudnick DA, He J, et al. Dyskerin ablation in mouse liver inhibits rRNA processing and cell division. *Mol Cell Biol.* 2010;30:413–422.

144. Raynaud CM, Sabatier L, Philipot O, Olaussen KA, Soria JC. Telomere length, telomeric proteins and genomic instability during the multistep carcinogenic process. *Crit Rev Oncol Hematol.* 2008;66:99–117.
145. Streffer C. Strong association between cancer and genomic instability. *Radiat Environ Biophys.* 2010;49:125–131.
146. de Lange T. Protection of mammalian telomeres. *Oncogene.* 2002;21:532–540.
147. de Lange T. Telomere-related genome instability in cancer. *Cold Spring Harb Symp Quant Biol.* 2005;70:197–204.
148. Sloand EM, Calado RT, Shah S. Telomere shortening and genomic instability: primary cells from patients with telomere repair complex mutations are susceptible to end-to-end chromo- some fusion and aneuploidy. *Blood.* 2006;108:2079.
149. Rampazzo E, Bonaldi L, Trentin L, et al. Telomere length and telomerase levels delineate subgroups of B-cell chronic lymphocytic leukemia with different biological characteristics and clinical outcomes. *Haematologica.* 2012;97:56–63.
150. Calado RT, Regal JA, Hills M, et al. Young NS Constitutional loss-of-function mutations in telomerase are genetic risk factors for acute myeloid leukemia. *Proc Natl Acad Sci USA.* 2009;106:1187–1192.
151. Hills M, Lücke K, Chavez EA, Eaves CJ, Lansdorp PM. Probing the mitotic history and developmental stage of hematopoietic cells using single telomere length analysis (STELA). *Blood.* 2009;113:5765–5775.
152. Scheinberg P, Nunez O, Weinstein B, et al. Horse versus rabbit antithymocyte globulin in acquired aplastic anemia. *N Engl J Med.* 2011;365:430–438.
153. Ramsay AJ, Quesada V, Foronda M, et al. POT1 mutations cause telomere dysfunction in chronic lymphocytic leukemia. *Nat Genet.* 2013;45:526–530.
154. Lin TT, Letsolo BT, Jones RE, et al. Telomere dysfunction and fusion during the progression of chronic lymphocytic leukemia: evidence for a telomere crisis. *Blood.* 2010;116:1899–1907.
155. Willeit P, Willeit J, Mayr A, et al. Telomere length and risk of incident cancer and cancer mortality. *JAMA.* 2010;304:69–75.
156. Cowan RW, Maitra A. Genetic progression of pancreatic cancer. *Cancer J.* 2014;20:80–84.
157. Ramakrishna G, Rastogi A, Trehanpati N, Sen B, Khosla R, Sarin SK. From cirrhosis to hepatocellular carcinoma: new molecular insights on inflammation and cellular senescence. *Liver Cancer.* 2013;2:367–383.

CHAPTER SEVEN

The Role of Telomeres in Liver Disease

Yoshiaki Sunami*, Guido von Figura†, Alexander Kleger‡, Pavel Strnad§, Norbert Hüser*, Daniel Hartmann*

*Department of General Surgery, Klinikum rechts der Isar, Technical University Munich, Munich, Germany
†Department of Internal Medicine II, Klinikum rechts der Isar, Technical University Munich, Munich, Germany
‡Department of Internal Medicine I, Ulm University Hospital, Ulm, Germany
§Department of Internal Medicine III and IZKF, University Hospital Aachen, Aachen, Germany

Contents

Abstract

Telomeres stabilize open chromosome ends and protect them against chromosomal end-to-end fusions, breakage, instability, and nonreciprocal translocations. Telomere dysfunction is known to lead to an impaired regenerative capacity of hepatocytes and an increased cirrhosis formation in the context of acute and chronic liver injury. In addition, telomere dysfunction and telomerase mutations have been associated with the induction of chromosomal instability and consequently with cirrhosis development and hepatocarcinogenesis. The identification of molecular mechanisms related to telomere dysfunction and telomerase activation might lead to new therapeutic strategies. In this chapter, we are reviewing the current knowledge about the importance of telomere dysfunction in liver diseases.

Progress in Molecular Biology and Translational Science, Volume 125
ISSN 1877-1173
http://dx.doi.org/10.1016/B978-0-12-397898-1.00007-4

1. TELOMERES AND TELOMERASE, AN OVERVIEW

Within eukaryotic cells, telomeres are located at the ends of linear chromosomes. They consist of conserved, noncoding tandem repeat nucleotide sequences (5′-TTAGGGn-3′)[1] and in humans extend each chromosome by about 10,000 to 15,000 additional nucleotides.[2,3] The major function of telomeres is to cap and stabilize open chromosome ends in order to prevent chromosomal end-to-end fusions, breakage, instability, and nonreciprocal translocations.[4,5]

In case of telomere dysfunction, free single-stranded chromosomal ends are recognized and fused by DNA ligase. During mitosis, these bonded chromosomes are disrupted thereby generating chromosome fragments that do not necessarily correspond to the original pieces.[6] These breakage–fusion–bridge cycles induce chromosomal instability. To avoid that, mammalian telomeres and associated proteins form the "shelterin complex," a multiprotein complex that plays an important role in protecting telomeres.[7,8] This complex is responsible for the tertiary structure of telomeres and consists of the six core factors TRF1/TERF1 (repeat-binding factor 1), TRF2/TERF2 (repeat-binding factor 2), TIN2/TINF2 (TERF1-interacting nuclear factor 2), POT1 (protection of telomere 1), TPP1/ACD (POT1- and TINF2-interacting protein), and RAP1 (transcription repressor/activator protein 1).[7,9] Mutations in telomere-binding proteins may impair the function of telomeres leading to defective telomere maintenance and the evolution of diseases.

Telomeres shorten by about 50–100 base pairs during each round of cell division due to the processing of telomeres during S phase of the cell cycle.[10,11] This process, also known as the "end-replication problem," is due to the fact that the DNA polymerase is unable to fully replicate the 3′ end of chromosomes.[10–12] Excessive telomere shortening leads to telomere dysfunction since the tertiary structure required for their capping cannot be formed anymore.[13] This is why the proliferative capacity of most human cells is limited to around 50–70 cell divisions.[13]

Telomere dysfunction is followed by the induction of DNA damage signaling cascades and a p53- and Rb-dependent permanent cell cycle arrest, named "replicative senescence".[14,15] The induction of senescence in response to telomere dysfunction can be prevented by deletion of p53/Rb, which in turn precludes the activation of p21 (WAF1/cyclin-dependent kinase inhibitor 1), a major cell cycle regulator.[15–17] An ablation of p53 leads to further telomere shortening and subsequently to an entry into a second, p53-independent checkpoint, called "crisis," that is characterized by apoptosis and chromosomal instability.[13,15,16] Both checkpoints constitute important tumor suppressors since they limit cell proliferation and thereby prevent immortal growth of tumor cells.[18] A schematic diagram of the telomere hypothesis is depicted in Fig. 7.1.

2. TELOMERE SHORTENING AND DISEASE PROGRESSION

Originally discovered by Carol Greider and Elizabeth Blackburn in the ciliated protozoan *Tetrahymena thermophila*, the evolutionary conserved holoenzyme telomerase is able to *de novo* synthesize 5′-TTAGGG-3′ hexameric telomere repeats and to add them catalytically onto the 3′-hydroxyl ends of human chromosomes.[18,21–23] The telomerase ribonucleoprotein enzyme complex is composed of the two essential components TERC (telomerase RNA component) and TERT (telomerase reverse transcriptase) as well as several associated proteins, such as dyskerin, GAR, NHP2, and NOP10.[18,24,25] TERT uses TERC as a template for telomeric repeat synthesis. While *TERC* is ubiquitously expressed in humans, the postnatal expression of *TERT* is restricted to germ cells or stem/progenitor cells of self-renewing organs, such as the liver, and is undetectable in most differentiated somatic tissues.[26,27]

While cells with high proliferative capacity need telomerase to counter telomere attrition,[22,28] its expression has to be strictly regulated in order to

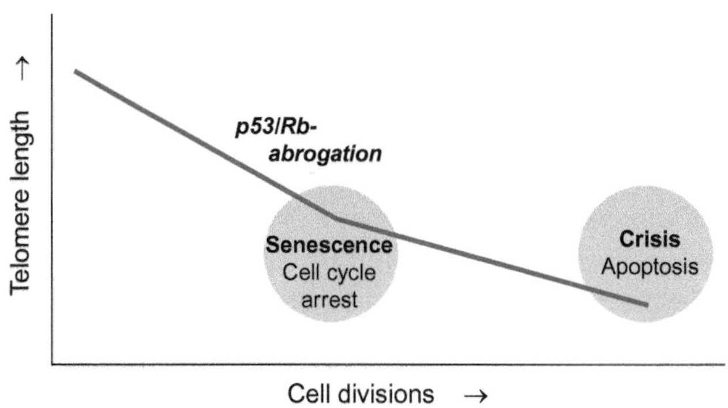

Figure 7.1 Telomere hypothesis. Telomere shortening limits the proliferative capacity of somatic cells and results in a permanent cell cycle arrest ("senescence") or apoptosis ("crisis"). *Adapted from Refs. 12,13,15,16,19,20.*

prevent the immortal growth of transformed cells.[29,30] Several components of the telomerase enzyme complex, such as dyskerin, have been shown to modulate telomerase activity and are responsible for the fact that gene defects cause accelerated telomere shortening.[31]

In the past years, several groups have generated *Terc*- or *Tert*-deficient animals in order to further elucidate the *in vivo* function of telomerase.[32–34] These studies revealed that a lack of telomerase activity in *Terc* knockout animals does not necessarily lead to a pathological phenotype. However, in successive generations of *Terc* knockout mice (from the third G3 to the sixth G6 generation), telomere dysfunction and chromosomal instability became more prominent.[32] These animals displayed accelerated aging, shortened life span, reduced tissue-repair abilities, and an impaired liver regeneration capacity.[36] Furthermore, cells derived from these animals showed malignant transformation and tumor formation.[35] The poor organ regeneration capacity in late-generation *Terc* knockout mice is due to the inhibition of cell cycle reentry of hepatocytes by the induction of senescence.[38] Telomere dysfunction in telomerase-deficient mice leads to an accumulation of DNA breaks by the induction of chromosomal fusions and breakages and the diminished regenerative potential causes the progression of chronic diseases, including development of liver cirrhosis.[35–37]

3. THE ROLE OF TELOMERE DYSFUNCTION IN CIRRHOSIS FORMATION

Liver cirrhosis is defined as an advanced stage of liver fibrosis characterized by the formation of regenerative nodules that are separated and encapsulated by fibrotic septa.[39] Cirrhosis development is known to be a consequence of repetitive liver tissue damage and can caused by triggers, such as toxic-/drug-induced (mainly alcohol), metabolic (mainly non-alcoholic fatty liver disease), infectious (mainly hepatitis B/C (HBC/HCV infection)), autoimmune (mainly primary biliary cirrhosis or autoimmune hepatitis), and others (biliary obstruction, vascular, and cryptogenic).[40] Liver cirrhosis represents the end stage of several chronic liver diseases and may lead to progressive hepatic failure and development of portal hypertension.[39] Furthermore, liver cirrhosis is known to be the main risk factor for the development of hepatocellular carcinoma (HCC).[39]

With respect to telomeres, two alternative models of cirrhosis formation and loss of organ function are being described in the literature.[40]

The first hypothesis assumes that chronic liver damage increases the rate of hepatic cell turnover leading to hepatocyte telomere shortening and regenerative exhaustion.[47] In agreement with this hypothesis, telomeres are significantly shorter in cirrhotic patients compared with noncirrhotic patients independent of the primary etiology and age and telomere shortening has, thus, been described as a general marker of cirrhosis formation.[48] In fibrotic areas, telomere shortening is restricted to hepatocytes, not affecting lymphocytes and stellate cells, and hepatocyte senescence correlates with the progression of fibrosis.[36,43] Telomere dysfunction and the subsequent induction of DNA damage pathways, including p21, are thought to contribute to the progression of chronic liver disease and to cirrhosis formation.[44–46] Despite that, it remains unclear, whether telomere shortening is the cause or the consequence of cirrhosis formation induced by chronic liver damage.[46]

The second hypothesis states that liver cirrhosis is a consequence of a wound-healing response with chronic inflammation and necrosis resulting in fibrotic scarring due to progressive collagen and matrix protein deposition.[41] According to this hypothesis, acute or chronic hepatic injury causes transformation of hepatic stellate cells (HSCs) from a quiescent to an activated state and leads to a production of extracellular matrix proteins (ECMs), such as collagen or smooth muscle actin (SMA).[42] ECM accumulation forms a fibrous scar and gradually distorts the hepatic architecture leading to the development of nodules of regenerating hepatocytes.[42,49] In this hypothesis, the interplay between telomere dysfunction and hepatic stellate cell activation remains to be elucidated. It is conceivable that telomere dysfunction triggers an inflammatory response, which activates hepatic stellate cells in a telomerase activity-independent manner.

4. ASSOCIATION OF TELOMERASE MUTATIONS WITH LIVER CIRRHOSIS

The development of liver dysfunction in cirrhotics is highly variable.[40,50] Moreover, around 5% of patients develop liver cirrhosis of unknown etiology.[43,51] This observation cannot be completely explained by environmental or demographic factors, and this fact led to the hypothesis that genetic predispositions affect the progression of liver cirrhosis.[48] In that respect, telomerase emerged as a potential disease modifier gene. For example, loss-of-function germline telomerase mutations were identified as the underlying cause of some human diseases, such as idiopathic pulmonary fibrosis.[52] These mutations may lead to a decrease in the stability of the telomerase complex or to a reduction of the enzymatic activity of telomerase.[48]

Two independent studies analyzed the prevalence of telomerase mutations and their influence on the development of cirrhosis in the context of chronic liver disease.[28,43] In the first of them, liver cirrhotics from different etiological backgrounds (mainly alcohol-induced, hepatitis B- or C-induced) were analyzed by large-scale sequencing of the telomerase components TERC and TERT. Sequencing analysis revealed an increased rate of telomerase mutation in patients with liver cirrhosis compared to healthy controls and HCV patients who did not develop cirrhosis.[43] *In vitro* studies of the identified cirrhosis-associated telomerase mutations revealed reduced telomerase activities and impaired telomere maintenance mechanisms. Cell lines carrying such mutations displayed reduced growth rates, and the mutation carriers had significantly decreased telomere lengths in their peripheral blood leukocytes.[28,43] These studies strongly suggest that telomerase dysfunction predisposes to cirrhosis development in response to chronic liver damage.[28,43]

To understand the underlying molecular mechanisms of liver fibrogenesis *in vivo*, several experimental fibrosis models have been established. Some of the most widely used models rely on repeated administration of carbon tetrachloride (CCl_4) or thioacetamide (TTA) or induction of chronic cholestasis via surgical bile duct ligation. Multiple injections of CCl_4 generate highly reactive free radicals, causing severe centrilobular necrosis and consequently liver fibrosis.[76] The role of telomerase in this widely used model has been analyzed *in vivo*. While steatosis and minimal fibrosis were observed in wild-type and G3 *Terc* knockout mice, G6 *Terc* knockout mice displayed a more pronounced steatosis and centrilobular fibrosis.[36] A reconstitution experiment with the adenoviral delivery of the *Terc* gene attenuated inflammation and the development of fibrosis in G6 *Terc* knockout mice, suggesting that the activation of telomerase during liver regeneration protects from the progression of liver fibrosis.[36]

5. THE ROLE OF TELOMERES IN METABOLIC LIVER DISEASE

Metabolic diseases, in particular obesity and diabetes mellitus type 2, are known to be associated with development of chronic liver diseases and constitute a risk factor for progression to liver cirrhosis.[57] Telomere shortening in peripheral blood cells has been shown to correlate with weight gain and an increased body mass index (BMI).[77] These data suggest that telomere shortening may have influence on early stages of liver diseases in patients with metabolic disorders.

G4 Terc-deficient mice display an impaired glucose tolerance and a reduced size of insulin-producing pancreatic beta cells, which suggests that telomere dysfunction leads to an impairment of pancreatic beta cell regeneration.[78,79] Under high-fat diet conditions, *Terc* knockout mice (G4) show an impaired insulin tolerance and an overexpression of p53 in adipose tissue.[80] Furthermore, the PPARγ coactivator 1α (PGC1α), a critical regulator of lipid metabolism, has been found to be downregulated.[81] Deficiency of Rap1, a component of the shelterin complex, causes mice to develop steatosis, impaired glucose tolerance, and increased abdominal fat depots, that is, features similar to what is found patients with metabolic problems.[82,83] However, the exact role of Rap1 deficiency in this phenotype as well as the association between telomere dysfunction and metabolism needs further clarification.

6. THE ROLE OF TELOMERES DURING CARCINOGENESIS

Telomerase activation is known to be the most common pathway leading to cellular immortalization and transformation.[58] Tumors have to activate telomere maintenance mechanisms, such as the reactivation of telomerase or alternative lengthening of telomeres (ALT), in order to obtain immortal growth capacity[8] and to bypass the tumor suppressor checkpoints "senescence" and "crisis." In agreement with these data, the ectopic expression of TERT in combination with the oncogenic allele of H-ras and the simian virus 40 large T oncoprotein represents one of the most powerful strategies to transform normal human epithelial and fibroblast cells into tumor cells.[59]

Several studies were able to show that genetic deletion or inhibition of telomerase leads to a significant suppression of the tumor-forming capacity of human cancer cells and to an attenuated tumor growth in mouse models.[54,60–62] In line with that, at least 85% of human cancer cells constitutively express TERT that confers them an unlimited growth potential via telomerase-dependent telomere elongation.[18,58,63,64] A schematic diagram of the role of telomeres and telomerase during carcinogenesis is depicted in Fig. 7.2.

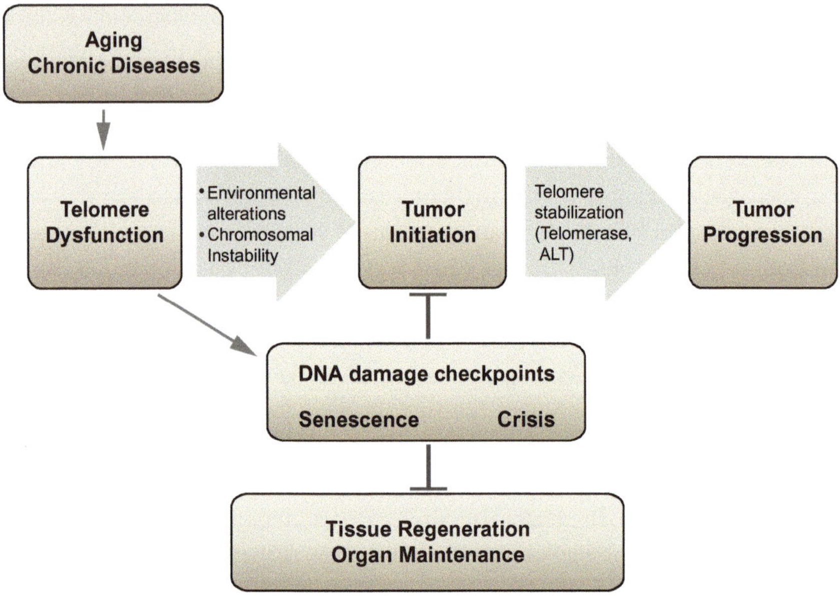

Figure 7.2 Role of telomeres during carcinogenesis. Telomere shortening in aged or chronically damaged tissue causes telomere dysfunction and increases the risk of cancer initiation.[40] To prevent the activation of DNA damage checkpoints and to become immortalized, tumor cells need to stabilize their telomeres either by activation of telomerase or by ALT.[40] *Adapted from Ref. 40.*

7. CHROMOSOMAL INSTABILITY AND HEPATOCARCINOGENESIS

HCC is the fifth most common cancer and the third most common cause of cancer mortality.[47,49] Telomere shortening is known to correlate with chromosomal instability, increased aneuploidy of chromosome 8, and the development of HCC in patients with different liver disease etiologies, such as hepatitis B and C infections and alcoholism.[69,70] Chromosomal instability has long been proposed to contribute to carcinogenesis, and in more than 90% of HCCs, instability of chromosomes 1 and 8 is observed.[84] Since gains at chromosome 8q22–24 correlate with the overexpression of the *MYC* gene in humans, it has been suggested that chromosomal instability might induce hepatocarcinogenesis via the overexpression of the oncogene c-myc (product of the *MYC* gene).[65–67] C-myc is known to be an important transactivator of the human TERT promoter and, thus, influences telomerase activity by activation of the catalytic subunit hTERT.[67] Accordingly, an increased c-myc expression has been observed in HCC mouse models and human tumor samples, and liver-specific c-myc expression in mice has been shown to promote hepatocarcinogenesis.[47,68] Gains at chromosome 15, which include a translocation of the *Myc* locus, have also been observed in human HCC samples.[37]

Telomerase activity is required for tumorigenesis and has been observed in more than 80% of tumors, including liver tumors.[72–75] Its importance was further supported by a study using telomerase inhibitors. Treatment with the thio-phosphoramidate oligonucleotide inhibitors GRN163 and GRN163L inhibited tumor growth and increased chemosensitivity of human HCC-derived cells.[29] The direct *in vivo* role of telomerase during carcinogenesis was demonstrated in a transgenic mouse model expressing hepatitis B surface antigen (HBsAg). The expression of HBsAg transgene causes hepatocellular injury, inflammation, and regenerative hyperplasia and ultimately leads to the development of HCC at the age of 18–21 months.[53] However, HBsAg transgene expression in combination with Terc deficiency reduced liver tumor development.[71]

Similarly, a combination of transgenic HBsAg overexpression with hepatic *p53*-deficiency leads to an attenuation of hepatocarcinogenesis when telomerase activity is depleted.[54] On the other hand, a deletion of p53 accelerates carcinogenesis in Terc knockout mice, and, when applying the CCl_4 injection-mediated liver injury model in G3/G4 Terc-deficient mice, loss of p53 heterozygosity leads to the development of advanced HCC.[85] As a potential mechanism contributing to an increased hepatocarcinogenesis, HCC-derived cells acquire replicative immortality by an inactivation of p53 and *CDKN2A* (p16INK) genes and by activation of *TERT* gene expression.[86] The fact that an inactivation of cell cycle checkpoints poses a critical step for a malignant transformation of hepatocytes is further supported by the loss of the expression of the senescence inducers p21 and p16 that is often found in HCCs with critical telomere shortening.[55] Of note, the downregulation of p21 that is required for HCC progression is mediated via the overexpression of Myc that in turn happens as a consequence of chromosomal instability (see above for details).[56]

8. TELOMERE DYSFUNCTION PROMOTES HEPATOCARCINOGENESIS

The current concept indicates that telomere dysfunction and telomerase reactivation occur sequentially during carcinogenesis. First, a telomere shortening due to aging or chronic disease induces chromosomal instability, and later on, the early cancer cells reactivate telomerase for genome stabilization and immortalization. This concept has recently been tested *in vivo* by the expression of a dominant-negative truncation of TRF2 (a member of "shelterin complex") in a telomerase-proficient background after carcinogen treatment.[37] It has been shown that transient telomere dysfunction induces chromosomal instability in tumor cells and increases hepatocarcinogenesis in telomerase-proficient mice. Telomerase promoted the progression of these chromosomally instable tumors by limiting DNA damage, cell cycle arrest, and apoptosis.[37] These data suggest that telomerase activation is required to stabilize tumors by chromosomal capping of ends with critically short telomeres thereby preventing the accumulation of DNA damage and subsequent checkpoint induction.[37]

9. CONCLUSION

In the context of liver injury, telomere shortening leads to an impaired regenerative capacity of hepatocytes and an increased cirrhosis development. In agreement with that, telomere dysfunction and telomerase mutations have been associated with the chromosomal instability, hepatocarcinogenesis, and progression of liver cirrhosis. Therefore, future studies should evaluate the usefulness of telomere-based indicators as prognostic markers for cirrhosis/tumor development. Given the crucial importance of telomeres in liver homeostasis, a better understanding of telomere biology might translate into new therapeutic strategies for patients with chronic liver diseases.

CONFLICT OF INTERESTS

The authors declare no potential conflicts of interests.

REFERENCES

1. Moyzis RK, Buckingham JM, Cram LS, et al. A highly conserved repetitive DNA sequence, (TTAGGG)n, present at the telomeres of human chromosomes. *Proc Natl Acad Sci U S A*. 1988;85(18):6622–6626.
2. Aviv A, Levy D, Mangel M. Growth, telomere dynamics and successful and unsuccessful human aging. *Mech Ageing Dev*. 2003;124(7):829–837.
3. Liu Y, Kha H, Ungrin M, Robinson MO, Harrington L. Preferential maintenance of critically short telomeres in mammalian cells heterozygous for mTert. *Proc Natl Acad Sci U S A*. 2002;99(6):3597–3602.
4. Counter CM, Avilion AA, LeFeuvre CE, et al. Telomere shortening associated with chromosome instability is arrested in immortal cells which express telomerase activity. *EMBO J*. 1992;11(5):1921–1929.

5. Hande MP, Samper E, Lansdorp P, Blasco MA. Telomere length dynamics and chromosomal instability in cells derived from telomerase null mice. *J Cell Biol.* 1999;144(4):589–601.
6. Murnane JP, Sabatier L. Chromosome rearrangements resulting from telomere dysfunction and their role in cancer. *BioEssays.* 2004;26(11):1164–1174.
7. de Lange T. Shelterin: the protein complex that shapes and safeguards human telomeres. *Genes Dev.* 2005;19(18):2100–2110.
8. Satyanarayana A, Manns MP, Rudolph KL. Telomeres, telomerase and cancer: an endless search to target the ends. *Cell Cycle.* 2004;3(9):1138–1150.
9. Liu D, O'Connor MS, Qin J, Songyang Z. Telosome, a mammalian telomere-associated complex formed by multiple telomeric proteins. *J Biol Chem.* 2004;279(49):51338–51342.
10. Kim NW, Piatyszek MA, Prowse KR, et al. Specific association of human telomerase activity with immortal cells and cancer. *Science.* 1994;266(5193):2011–2015.
11. Lundblad V. The end replication problem: more than one solution. *Nat Med.* 1997;3(11):1198–1199.
12. Hayflick L. Human cells and aging. *Sci Am.* 1968;218(3):32–37.
13. Allsopp RC, Vaziri H, Patterson C, et al. Telomere length predicts replicative capacity of human fibroblasts. *Proc Natl Acad Sci U S A.* 1992;89(21):10114–10118.
14. d'Adda di Fagagna F, Reaper PM, Clay-Farrace L, et al. A DNA damage checkpoint response in telomere-initiated senescence. *Nature.* 2003;426(6963):194–198.
15. Wright WE, Shay JW. The two-stage mechanism controlling cellular senescence and immortalization. *Exp Gerontol.* 1992;27(4):383–389.
16. Artandi SE, DePinho RA. A critical role for telomeres in suppressing and facilitating carcinogenesis. *Curr Opin Genet Dev.* 2000;10(1):39–46.
17. Brown JP, Wei W, Sedivy JM. Bypass of senescence after disruption of p21CIP1/WAF1 gene in normal diploid human fibroblasts. *Science.* 1997;277(5327):831–834.
18. Calado RT, Young NS. Telomere diseases. *New Eng J Med.* 2009;361(24):2353–2365.
19. Bodnar AG, Ouellette M, Frolkis M, et al. Extension of life-span by introduction of telomerase into normal human cells. *Science.* 1998;279(5349):349–352.
20. Vaziri H, Dragowska W, Allsopp RC, Thomas TE, Harley CB, Lansdorp PM. Evidence for a mitotic clock in human hematopoietic stem cells: loss of telomeric DNA with age. *Proc Natl Acad Sci U S A.* 1994;91(21):9857–9860.
21. Cohen SB, Graham ME, Lovrecz GO, Bache N, Robinson PJ, Reddel RR. Protein composition of catalytically active human telomerase from immortal cells. *Science.* 2007;315(5820):1850–1853.
22. Greider CW, Blackburn EH. Identification of a specific telomere terminal transferase activity in Tetrahymena extracts. *Cell.* 1985;43(2 Pt 1):405–413.
23. Nugent CI, Lundblad V. The telomerase reverse transcriptase: components and regulation. *Genes Dev.* 1998;12(8):1073–1085.
24. Greider CW, Blackburn EH. A telomeric sequence in the RNA of Tetrahymena telomerase required for telomere repeat synthesis. *Nature.* 1989;337(6205):331–337.
25. Meyerson M, Counter CM, Eaton EN, et al. hEST2, the putative human telomerase catalytic subunit gene, is up-regulated in tumor cells and during immortalization. *Cell.* 1997;90(4):785–795.
26. Collins K, Mitchell JR. Telomerase in the human organism. *Oncogene.* 2002;21(4):564–579.
27. Wright WE, Piatyszek MA, Rainey WE, Byrd W, Shay JW. Telomerase activity in human germline and embryonic tissues and cells. *Dev Genet.* 1996;18(2):173–179.
28. Calado RT, Brudno J, Mehta P, et al. Constitutional telomerase mutations are genetic risk factors for cirrhosis. *Hepatology.* 2011;53(5):1600–1607.

29. Djojosubroto MW, Chin AC, Go N, et al. Telomerase antagonists GRN163 and GRN163L inhibit tumor growth and increase chemosensitivity of human hepatoma. *Hepatology*. 2005;42(5):1127–1136.
30. Wright WE, Shay JW. Cellular senescence as a tumor-protection mechanism: the essential role of counting. *Curr Opin Genet Dev*. 2001;11(1):98–103.
31. Mitchell JR, Wood E, Collins K. A telomerase component is defective in the human disease dyskeratosis congenita. *Nature*. 1999;402(6761):551–555.
32. Blasco MA, Lee HW, Hande MP, et al. Telomere shortening and tumor formation by mouse cells lacking telomerase RNA. *Cell*. 1997;91(1):25–34.
33. Chiang YJ, Hemann MT, Hathcock KS, et al. Expression of telomerase RNA template, but not telomerase reverse transcriptase, is limiting for telomere length maintenance in vivo. *Mol Cell Biol*. 2004;24(16):7024–7031.
34. Yuan X, Ishibashi S, Hatakeyama S, et al. Presence of telomeric G-strand tails in the telomerase catalytic subunit TERT knockout mice. *Genes Cells*. 1999;4(10):563–572.
35. Rudolph KL, Chang S, Lee HW, et al. Longevity, stress response, and cancer in aging telomerase-deficient mice. *Cell*. 1999;96(5):701–712.
36. Rudolph KL, Chang S, Millard M, Schreiber-Agus N, DePinho RA. Inhibition of experimental liver cirrhosis in mice by telomerase gene delivery. *Science*. 2000;287(5456):1253–1258.
37. Begus-Nahrmann Y, Hartmann D, Kraus J, et al. Transient telomere dysfunction induces chromosomal instability and promotes carcinogenesis. *J Clin Invest*. 2012;122(6):2283–2288.
38. Satyanarayana A, Wiemann SU, Buer J, et al. Telomere shortening impairs organ regeneration by inhibiting cell cycle re-entry of a subpopulation of cells. *EMBO J*. 2003;22(15):4003–4013.
39. Pinzani M, Macias-Barragan J. Update on the pathophysiology of liver fibrosis. *Exp Rev Gastroenterol Hepatol*. 2010;4(4):459–472.
40. Rudolph KL, Hartmann D, Opitz OG. Telomere dysfunction and DNA damage checkpoints in diseases and cancer of the gastrointestinal tract. *Gastroenterology*. 2009;137(3):754–762.
41. Friedman SL. Hepatic stellate cells: protean, multifunctional, and enigmatic cells of the liver. *Physiol Rev*. 2008;88(1):125–172.
42. Bataller R, Brenner DA. Liver fibrosis. *J Clin Invest*. 2005;115(2):209–218.
43. Hartmann D, Srivastava U, Thaler M, et al. Telomerase gene mutations are associated with cirrhosis formation. *Hepatology*. 2011;53(5):1608–1617.
44. Aikata H, Takaishi H, Kawakami Y, et al. Telomere reduction in human liver tissues with age and chronic inflammation. *Exp Cell Res*. 2000;256(2):578–582.
45. Kitada T, Seki S, Kawakita N, Kuroki T, Monna T. Telomere shortening in chronic liver diseases. *Biochem Biophys Res Commun*. 1995;211(1):33–39.
46. Wiemann SU, Satyanarayana A, Tsahuridu M, et al. Hepatocyte telomere shortening and senescence are general markers of human liver cirrhosis. *FASEB J*. 2002;16(9):935–942.
47. El-Serag HB, Rudolph KL. Hepatocellular carcinoma: epidemiology and molecular carcinogenesis. *Gastroenterology*. 2007;132(7):2557–2576.
48. Chaiteerakij R, Roberts LR. Telomerase mutation: a genetic risk factor for cirrhosis. *Hepatology*. 2011;53(5):1430–1432.
49. Williams EJ, Iredale JP. Liver cirrhosis. *Postgrad Med J*. 1998;74(870):193–202.
50. Lechel A, Manns MP, Rudolph KL. Telomeres and telomerase: new targets for the treatment of liver cirrhosis and hepatocellular carcinoma. *J Hepatol*. 2004;41(3):491–497.
51. Kodali VP, Gordon SC, Silverman AL, McCray DG. Cryptogenic liver disease in the United States: further evidence for non-A, non-B, and non-C hepatitis. *Am J Gastroenterol*. 1994;89(10):1836–1839.

52. Armanios MY, Chen JJ, Cogan JD, et al. Telomerase mutations in families with idiopathic pulmonary fibrosis. *New Eng J Med*. 2007;356(13):1317–1326.
53. Chisari FV, Klopchin K, Moriyama T, et al. Molecular pathogenesis of hepatocellular carcinoma in hepatitis B virus transgenic mice. *Cell*. 1989;59(6):1145–1156.
54. Lechel A, Holstege H, Begus Y, et al. Telomerase deletion limits progression of p53-mutant hepatocellular carcinoma with short telomeres in chronic liver disease. *Gastroenterology*. 2007;132(4):1465–1475.
55. Plentz RR, Park YN, Lechel A, et al. Telomere shortening and inactivation of cell cycle checkpoints characterize human hepatocarcinogenesis. *Hepatology*. 2007;45(4):968–976.
56. Hui L, Zatloukal K, Scheuch H, Stepniak E, Wagner EF. Proliferation of human HCC cells and chemically induced mouse liver cancers requires JNK1-dependent p21 downregulation. *J Clin Invest*. 2008;118(12):3943–3953.
57. Farrell GC, Larter CZ. Nonalcoholic fatty liver disease: from steatosis to cirrhosis. *Hepatology*. 2006;43(2 Suppl 1):S99–S112.
58. Wyatt HD, West SC, Beattie TL. InTERTpreting telomerase structure and function. *Nucleic Acids Res*. 2010;38(17):5609–5622.
59. Hahn WC, Counter CM, Lundberg AS, Beijersbergen RL, Brooks MW, Weinberg RA. Creation of human tumour cells with defined genetic elements. *Nature*. 1999;400(6743):464–468.
60. Farazi PA, Glickman J, Jiang S, Yu A, Rudolph KL, DePinho RA. Differential impact of telomere dysfunction on initiation and progression of hepatocellular carcinoma. *Cancer Res*. 2003;63(16):5021–5027.
61. Gonzalez-Suarez E, Samper E, Flores JM, Blasco MA. Telomerase-deficient mice with short telomeres are resistant to skin tumorigenesis. *Nat Genet*. 2000;26(1):114–117.
62. Rudolph KL, Millard M, Bosenberg MW, DePinho RA. Telomere dysfunction and evolution of intestinal carcinoma in mice and humans. *Nat Genet*. 2001;28(2):155–159.
63. Hornsby PJ. Senescence and life span. *Pflugers Arch*. 2010;459(2):291–299.
64. Kim NW, Wu F. Advances in quantification and characterization of telomerase activity by the telomeric repeat amplification protocol (TRAP). *Nucleic Acids Res*. 1997;25(13):2595–2597.
65. Lengauer C, Kinzler KW, Vogelstein B. Genetic instabilities in human cancers. *Nature*. 1998;396(6712):643–649.
66. Thorgeirsson SS, Grisham JW. Molecular pathogenesis of human hepatocellular carcinoma. *Nat Genet*. 2002;31(4):339–346.
67. Wu KJ, Grandori C, Amacker M, et al. Direct activation of TERT transcription by c-MYC. *Nat Genet*. 1999;21(2):220–224.
68. Zender L, Kubicka S. Molecular pathogenesis and targeted therapy of hepatocellular carcinoma. *Onkologie*. 2008;31(10):550–555.
69. Plentz RR, Caselitz M, Bleck JS, et al. Hepatocellular telomere shortening correlates with chromosomal instability and the development of human hepatoma. *Hepatology*. 2004;40(1):80–86.
70. Plentz RR, Schlegelberger B, Flemming P, et al. Telomere shortening correlates with increasing aneuploidy of chromosome 8 in human hepatocellular carcinoma. *Hepatology*. 2005;42(3):522–526.
71. Wiemann SU, Satyanarayana A, Buer J, Kamino K, Manns MP, Rudolph KL. Contrasting effects of telomere shortening on organ homeostasis, tumor suppression, and survival during chronic liver damage. *Oncogene*. 2005;24(9):1501–1509.
72. Kojima H, Yokosuka O, Imazeki F, Saisho H, Omata M. Telomerase activity and telomere length in hepatocellular carcinoma and chronic liver disease. *Gastroenterology*. 1997;112(2):493–500.

73. Miura N, Horikawa I, Nishimoto A, et al. Progressive telomere shortening and telomerase reactivation during hepatocellular carcinogenesis. *Cancer Genet Cytogenet.* 1997;93(1):56–62.
74. Nakayama J, Tahara H, Tahara E, et al. Telomerase activation by hTRT in human normal fibroblasts and hepatocellular carcinomas. *Nat Genet.* 1998;18(1):65–68.
75. Suda T, Isokawa O, Aoyagi Y, et al. Quantitation of telomerase activity in hepatocellular carcinoma: a possible aid for a prediction of recurrent diseases in the remnant liver. *Hepatology.* 1998;27(2):402–406.
76. Tsukamoto H, Matsuoka M, French SW. Experimental models of hepatic fibrosis: a review. *Semin Liver Dis.* 1990;10(1):56–65.
77. Gardner JP, Li S, Srinivasan SR, et al. Rise in insulin resistance is associated with escalated telomere attrition. *Circulation.* 2005;111(17):2171–2177.
78. Liew CW, Holman A, Kulkarni RN. The roles of telomeres and telomerase in beta-cell regeneration. *Diabetes Obes Metabol.* 2009;11(suppl 4):21–29.
79. Kuhlow D, Florian S, von Figura G, et al. Telomerase deficiency impairs glucose metabolism and insulin secretion. *Aging (Albany NY).* 2010;2:650–658.
80. Minamino T, Orimo M, Shimizu I, et al. A crucial role for adipose tissue p53 in the regulation of insulin resistance. *Nat Med.* 2009;15(9):1082–1087.
81. Li X, Monks B, Ge Q, Birnbaum MJ. Akt/PKB regulates hepatic metabolism by directly inhibiting PGC-1alpha transcription coactivator. *Nature.* 2007;447(7147):1012–1016.
82. Yeung F, Ramírez CM, Mateos-Gomez PA, et al. Nontelomeric role for Rap1 in regulating metabolism and protecting against obesity. *Cell Rep.* 2013;3:1847–1856.
83. Martínez P, Gomez-López PA, García F, et al. RAP1 protects from obesity through its extratelomeric role regulating gene expression. *Cell Rep.* 2013;3:2059–2074.
84. Wilkens L, Flemming P, Gebel M, et al. Induction of aneuploidy by increasing chromosomal instability during dedifferentiation of hepatocellular carcinoma. *Proc Natl Acad Sci U S A.* 2004;101(5):1309–1314.
85. Farazi PA, Glickman J, Horner J, DePinho RA. Cooperative interactions of p53 mutation, telomere dysfunction, and chronic liver damage in hepatocellular carcinoma progression. *Cancer Res.* 2006;66(9):4766–4773.
86. Ozturk N, Erdal E, Mumcuoglu M, et al. Reprogramming of replicative senescence in hepatocellular carcinoma-derived cells. *Proc Natl Acad Sci U S A.* 2006;103(7):2178–2183.

CHAPTER EIGHT

Telomeres in Lung Diseases

Rodrigo T. Calado
Department of Internal Medicine, University of São Paulo at Ribeirão Preto School of Medicine, Ribeirão Preto, São Paulo, Brazil

Contents

Abstract

Idiopathic pulmonary fibrosis is a progressive and fatal interstitial lung disease leading to respiratory failure. Mutations in telomerase complex genes (TERT or TERC) and short telomeres are genetic risk factors for the development of familial or sporadic idiopathic pulmonary fibrosis. Up to 15% of familial cases and approximately 5% of sporadic cases carry a heterozygous mutation in one of the genes, and patients' cells retain approximately 50% of telomerase activity. Pulmonary fibrosis also is a manifestation of dyskeratosis congenita, an inherited bone marrow failure syndrome caused by telomere dysfunction. Short telomeres even in the absence of telomerase mutations are a feature of most patients with idiopathic pulmonary fibrosis. Telomerase mutations also have been linked to pulmonary fibrosis and emphysema syndrome. Although short telomeres have been clearly linked to idiopathic pulmonary fibrosis, the mechanisms of disease are still unclear.

1. INTRODUCTION

That telomere dysfunction may associate with lung disease was first observed in dyskeratosis congenita (DC): pulmonary fibrosis develops in up to 20% of children diagnosed with DC.[1] DC is an inherited bone marrow failure syndrome caused by mutations in telomere-biology genes and very short telomeres mainly affecting children[2] and discussed in another chapter of this book. Respiratory insufficiency also is a common fatal complication

Progress in Molecular Biology and Translational Science, Volume 125
ISSN 1877-1173
http://dx.doi.org/10.1016/B978-0-12-397898-1.00008-6

after hematopoietic stem cell transplant for the treatment of bone marrow failure in DC[3,4]; in some cases, pulmonary fibrosis may develop after hematopoietic stem cell transplant.[5] Using different investigational strategies, several groups have linked mutations in telomerase genes to familial and sporadic idiopathic pulmonary fibrosis. Telomere attrition also has been associated with combined pulmonary fibrosis and emphysema syndrome.[6,7] Telomerase-deficient animal models have been developed to address the contribution of abnormal telomere maintenance to lung fibrosis, with conflicting results. The mechanisms by which short telomeres cause pulmonary fibrosis are still poorly understood and may be due to defective tissue repair, inflammation, or a combination of genetic and environmental factors.

In this chapter, we discuss the role of telomere dysfunction to lung disorders and how it contributes to better understand disease mechanism.

2. IDIOPATHIC PULMONARY FIBROSIS

Idiopathic pulmonary fibrosis (IPF) is an age-related disorder characterized by chronic, progressive, and usually fatal fibrosing interstitial pneumonia. It is clinically manifested by cough and dyspnea of no apparent etiology, and clinical evaluation unravels typical radiological and histopathologic findings of usual interstitial pneumonia (UIP).[8,9] Radiological imaging, mainly high-resolution computed tomography (HRCT) scan, is crucial in the diagnosis of IPF and consistent with UIP (Fig. 8.1). It may display reticular opacity, ground glass opacity, and honeycombing. Lung involvement is often patchy, basal, and peripheral. Histologically, there is patchy involvement of the parenchyma with fibrosis, tissue distortion, honeycombing, and fibroblast foci.

IPF is a disease restricted to the lungs without the involvement of other organs or systems. In most patients, IPF worsens progressively, but some patients may experience acute worsening of respiratory symptoms, whereas in a minority of patients, IPF may be stable over the years. IPF may be sporadic (without family history) or, in a minority of cases (approximately 5%), familial, in which two or more family members are affected. Familial and sporadic IPF are not clinically or pathologically distinguishable, although familial cases tend to present at earlier ages.

The estimated incidence of IPF increases with age and is approximately 11 new cases in 100,000 individuals in men and 7 new cases in 100,000 individuals in women.[10] Although IPF is by definition idiopathic, there are risk factors associated with disease development, such as cigarette smoking;

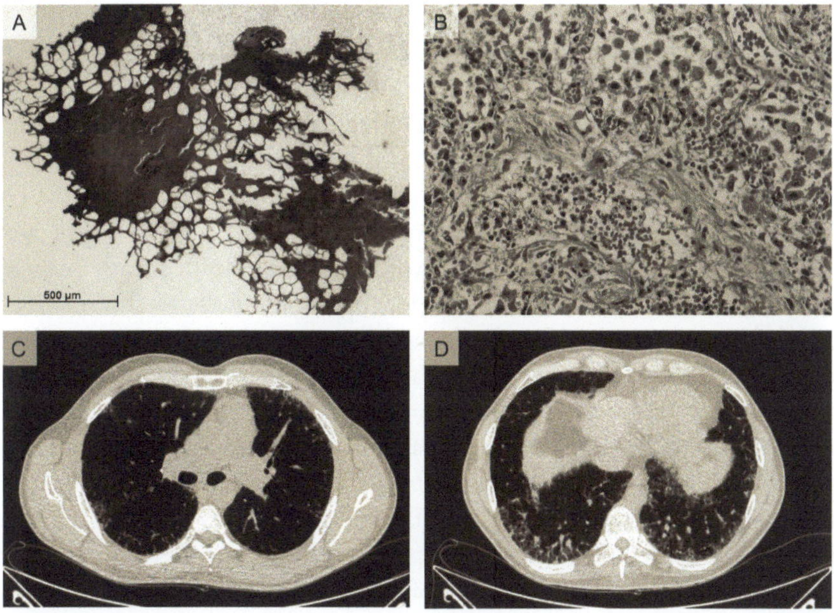

Figure 8.1 (A) Lung biopsy of a patient with idiopathic pulmonary fibrosis and a *TERT* mutation demonstrating extensive fibrosis. (B) Biopsy of another patient illustrating significant fibrosis and inflammation. (C and D) High-resolution computed tomography scans illustrating patchy peripheral lung fibrosis especially in the bases with honeycombing. (See the color plate.)

exposure to metal, wood, and animal dusts; stone polishing; hair dressing; Epstein–Barr virus; and gastroesophageal reflux.

There is no beneficial pharmacological therapy for IPF and lung transplant is the only potentially curative procedure.[8] Inflammation and chronic exposure to injury and repair appear to be part of the mechanism of disease.[9] However, anti-inflammatory drugs do not provide clinical response, challenging the contribution of inflammation to disease progression. Hyperplasia of alveolar epithelial type II cells (AECIIs) is observed in IPF along with abnormal secretion of profibrotic products.[11] Additionally, there are an increased number of fibroblasts in the lung tissue in IPF, which appears to contribute to tissue destruction. Fibroblasts form "fibroblast foci," in which excess collagen and extracellular matrix are found, producing thickened alveolar walls and distorting tissue architecture.[12]

Although most cases of IPF are sporadic, the most common genetic pattern of contribution to IPF appears to be autosomal dominant with variable inheritance. Several genes have been implicated as etiologic in IPF. The

ELMOD2 gene on chromosome 4q31 has been linked to familial IPF.[13] Stronger association with IPF has been observed with the mucin *MUC5B* gene[14] and surfactant C protein *SFTPC* gene.[15]

Using different genetic approaches, three independent groups implicated the *TERT* gene in IPF. Applying genome-wide association study (GWAS) in a cohort of 159 patients with IPF, a Japanese group found an association between a single-nucleotide polymorphism (SNP) in the *TERT* gene and IPF.[16] A second group performed whole genome SNP linkage scan in two large families with familial IPF, identifying a linkage to chromosome 5p15.[17] As the *TERT* gene is in the region and a potential link between telomerase and pulmonary fibrosis had been observed in 20% of patients with DC,[1] the candidate *TERT* gene was sequenced, unraveling rare nonsynonymous mutations. Affected patients had very short telomeres in peripheral blood leukocytes and *in vitro* functional experiments determined that *TERT* mutations found the two families significantly impaired telomerase reverse transcriptase enzymatic activity. Patients were heterozygous for the mutations, indicating reduction of 50% in enzyme function is sufficient for disease development. However, disease penetrance was variable. Of note, several relatives in these two families had some features of DC, such as anemia or cancer development, but they lacked the classical mucocutaneous triad of nail dystrophy, leukoplakia, and reticular skin hyperpigmentation, or overt bone marrow failure. The authors then sequenced the *TERT* gene in an additional 44 families with idiopathic lung disease and identified an additional four heterozygous *TERT* gene mutations, also modulating telomerase activity. They also identified a heterozygous *TERC* gene mutation, which encodes telomerase RNA component that provides the template for telomerase reverse transcriptase. To further investigate the role of telomerase mutations in IPF, they screened 44 cases of sporadic IPF and identified an additional heterozygous *TERT* mutation.

A third group identified a *TERT* mutation in a large family with DC, in which pulmonary fibrosis was a common feature.[18] Given this association, the authors screened 73 probands with IPF for mutations in *TERT* and *TERC* and identified six novel heterozygous mutations in one of the genes.[19] Patients also had short telomeres in peripheral blood leukocytes and mutations significantly reduced telomerase enzymatic activity. Again, asymptomatic family members carrying the mutation also had short telomeres, suggesting variable penetrance and that they may also be at risk for lung disease. That *TERT* mutation carriers have subclinical lung involvement has been demonstrated in the study of first-degree relatives of probands with familial IPF.[20] In clinically healthy *TERT* mutation carriers, lymphocyte infiltrate in bronchoalveolar lavage was observed 20 years before clinical IPF developed, in spite of normal computed tomography imaging. Other mutation carriers may remain clinically asymptomatic, but with sign of lung interstitial disease on HRCT scan.

Taken together, these studies firmly establish telomerase mutations as genetic risk factors for familial or sporadic IPF. Telomerase mutations are found in up to 15% of familial cases and in approximately 3% of sporadic cases.[21] In IPF patients, most mutations are heterozygous and in either the *TERT* or the *TERC* genes; affected individuals retain at least 50% of telomerase activity, which is not sufficient to maintain adequate telomere lengths. These findings parallel to what is observed in aplastic anemia, in which most mutations are in the two essential telomerase components. However, it contrasts with findings in DC, in which mutations may be heterozygous (autosomal dominant), biallelic (autosomal recessive), or hemizygous, covering a larger spectrum of genes involved. In DC, the essential telomerase complex genes (*TERT* and *TERC*) may be involved and also *DKC1*, in the X chromosome, which encodes a telomerase complex-associated protein, *NHP2*; *NOP1*, also encoding associate components of the telomerase complex; *TINF2*, which encodes the shelterin protein TIN2; *RTEL1*, which encodes a DNA helicase and disrupts D-loop formation during homologous recombination; *WRAP53*, which encodes TCAB1, a protein that localizes telomerase to the Cajal bodies; and CTC1, implicated in telomere maintenance. Mutations in these genes are very rarely found in IPF.[22] When these mutations occur, the usual phenotype appears in the first decade of life and pulmonary fibrosis is one of the many manifestations of DC.

In spite of IPF being a severe and fatal disease, along with aplastic anemia, it appears to be a "less severe" phenotype of telomere diseases in comparison with DC and its very severe variants[2] (Hoyeraal–Hreidarsson syndrome and Revesz syndrome): IPF manifests later in life, usually after the sixth decade of life, and usually affects only the lungs. In less than 5% of patients with IPF, the liver also is affected.[23] Whereas in DC, gene mutation is disease-determinant and severely affects telomerase function or telomere maintenance, patients have very short telomeres, and genetic penetrance is high, in IPF, gene lesion behaves as a risk factor, telomerase impairment is less intense, and other genetic, epigenetic, and environmental factors also play a role in disease progression.[23] In a series of 134 heterozygous *TERT* mutation carriers in 21 unrelated families, IPF was age-dependent and not observed in subjects younger than 40 years and was associated with exposure to fibrogenic exposure, such as cigarette smoking.[24] A minority of patients also presented cytopenias or liver cirrhosis.

The natural history of telomerase-deficient IPF appears to be comparable to those in which no mutation is observed. Most patients are affected after age 60 years, males are predominantly affected, and the median survival is 3 years after diagnosis.[24] In addition, while telomerase mutations are found in a minority of IPF patients, most of them have short telomeres for their age.[23] Short telomeres even in the absence of a recognizable mutation appear to contribute to IPF. The existence of telomeres without telomerase mutations is still not clear, but it may be explained by several factors. First, telomere length may be modulated by environmental factors, and cigarette smoking, dietary content, and lifestyle have been blamed to shorten telomeres in humans.[2] Second, patients are routinely screened for mutations in the coding or adjacent intronic regions of telomerase genes, but mutations in regulatory regions may decrease gene expression, as observed in patients with aplastic anemia.[25] Third, patients may carry mutations in telomere-biology genes yet to be discovered. Finally, the pattern of telomere length inheritance is complex; patients may inherit short telomeres from their mutant parent but not the mutation itself. Although they retain normal telomerase activity, they are born with shorter telomeres in comparison with healthy subjects. Normal telomerase activity does not "reset" telomere length to normal levels, and it may take several generations for telomeres to return to normal lengths.[26]

3. TELOMERES AND EMPHYSEMA

IPF with UIP pattern is not the only pulmonary manifestation of telomere diseases. In a minority of patients, chronic hypersensitivity pneumonitis and interstitial pneumonitis also have been reported.[24] Additionally, telomerase mutations appear to be the genetic mechanism for the combined pulmonary fibrosis and emphysema syndrome.[7] Telomerase "knockout" mice with short telomeres exposed to chronic cigarette smoke for 6 months developed emphysema with air space enlargement rather than fibrosis.[27] Telomerase-deficient mice exposed to cigarette smoke also developed decreased lung volume-adjusted compliance. Although it contrasts with what is usually observed in humans with emphysema, who develop increased compliance, it appears to be a characteristic of emphysema in rodents, and collagen deposition was not observed in these mice.[23] Emphysema in telomerase-null mice exposed to cigarette smoke also was a consequence of short telomeres in lung tissue only, as "knockout" mice transplanted with wild-type bone marrows also developed the same pulmonary pattern of enlarged air spaces. Cigarette smoke also induced DNA damage in lung endothelium and impaired proliferation of type II alveolar epithelial cells. In humans, telomerase mutations have been described in at least two families with either TERT or TERC mutations in which affected subjects presented either IPF or combined pulmonary fibrosis and emphysema syndrome.[7,27] Also in these families, cigarette smoking appeared to modulate disease.

4. TELOMERES AND DEVELOPMENT OF LUNG DISEASE

Although telomerase mutations and telomere erosion have been clearly associated with IPF, the mechanisms by which dysfunctional telomeres eventually lead to lung fibrosis are not still clear. For bone marrow failure and cancer development, the mechanisms are better understood. Short telomeres activate p53 and eventually signal proliferative senescence and/or apoptosis.[28] Hematopoietic stem and progenitor cells display a reduced stem cell pool and a deficient proliferative capacity, clinically resulting and a hypoplastic marrow and decreased ability to maintain peripheral blood cell numbers. The levels for growth factors, such as the stem cell factor, granulocyte colony-stimulating factor, erythropoietin, and thrombopoietin, are increased in patients' serum in response to deficient hematopoiesis. Dysfunctional telomeres also are more prone for rearrangement, recombination, or DNA damage, making cells more susceptible to malignant transformation.

For pulmonary cells, the mechanisms for disease development are less clear, and this is partly due to the fact that lung samples are difficult to obtain from patients and most lung epithelial cells are not present anymore by the time of clinical presentation. This combination makes the study of patients' samples very difficult for researchers to better understand disease mechanism.

Animal models have provided conflicting results and their contribution to the study of telomere disease has been limited.[29] *Tert*-deficient animals challenged with bleomycin, a chemotherapeutic agent that also induces lung fibrosis, presented significantly less pulmonary fibrosis in comparison with wild-type animals.[30] Lung samples from *Tert*-null mice exposed to bleomycin also expressed less α-SMA, a marker for myeloblast differentiation, decreased differentiation, and more apoptosis. Additionally, *Terc*-deficient animals did not show any modulation in bleomycin-induced lung fibrosis.[31] Telomerase expression also was activated in fibroblasts from fibrotic murine lungs induced by bleomycin. It is difficult to reconcile the findings in humans and mice. One explanation is that bleomycin-induced lung fibrosis is more acute and is dependent of mechanisms different from the chronic injury-repair pattern observed in IPF. Alternatively, the pathological mechanisms for pulmonary fibrosis in humans and rodents may be diverse and immunologically distinct.

In another study mentioned above, telomerase-null mice exposed to cigarette smoke developed emphysema but not fibrosis.[27] This mouse model highlights one of the pathological patterns observed in some patients with telomerase deficiency, namely, emphysema. Again, the lack of fibrosis may be due to the different inflammatory mechanisms activated in fibrosis in rodents. Additionally, the period of cigarette smoke exposure in this study was relatively short (6 months) in comparison with the long exposure observed in humans. IPF develops after decades of smoking and it is possible that the mouse model is not completely adequate to reproduce human disease.

Alveolar inflammation appears to be a very early event in the development of IPF in telomerase mutation carriers. Two daughters of a proband with biopsy-proven IPF were evaluated in the 1980s for lung function.[32] Although their lung function and radiological imaging were normal at the time, they presented alveolar inflammation and activated alveolar macrophages with high numbers of lymphocytes and neutrophils in their bronchoalveolar lavage. Thirty years later, they were evaluated and found to carry a *TERT* gene mutation.[20] They had very short telomeres and mutation reduced telomerase enzymatic activity by approximately 70%. Whereas one sister developed IPF, the other was clinically asymptomatic after 30 years, but her HRCT scan unraveled mild changes consistent with early IPF. Alveolar inflammation may be responsible for epithelial injury in the first stages of IPF.

Normal telomerase function and longer telomeres may also be necessary for appropriate lung regeneration and alveolar epithelium repair. Short telomeres may compromise cellular proliferative capacity or induce apoptosis.[27] It has been proposed that senescent fibrogenic cells in the injured liver may mediate antifibrotic effects[33] and senescent cells may be seen as fibrosis suppressors.[34] However, short telomeres increase the percentage of senescent cells, but the observed effect is profibrotic. It is possible that in the lungs, cell senescence represses the fibrotic response. In tissue samples of patients with IPF, telomerase expression is increased in fibroblasts,[31] suggesting that telomerase is necessary for the injury healing process. Lower telomerase activity may impair appropriate lung healing.

Another proposed mechanism for IPF is stem cell exhaustion, specifically of type II alveolar epithelium cells.[35] Eroded and dysfunctional telomeres impair adequate stem cell function leading to a reduced stem cell pool, which has been demonstrated for the hematopoietic tissue.[28,36,37] However, it is not clear whether short telomeres also provoke a depletion of type II alveolar epithelium cells.

In the group of telomere diseases, fibrosis occurs not only in the lungs but also in the liver. Cirrhosis is observed in some patients with aplastic anemia and manifest along with IPF in a minority of patients.[1,23,38] The lungs and the liver share histopathologic findings in telomere diseases: fibrosis and inflammation.[39] However, the mechanisms underlying these findings in the liver also are unknown. It is likely that these organs share similar regenerative and profibrotic mechanisms that are abnormal in subjects with short telomeres.

Since normal lung cells are not present at the time of lung biopsy, the study of affected cells is troublesome. Animal models have not proven to elucidate the molecular and cellular pathways involved in IPF. Reprogramming of adult patient-specific cells is a novel tool in translational medicine. Induced pluripotent stem (iPS) cells have been derived from patients with telomere diseases.[40–42] These cells may be an important tool for the research of IPF in the setting of telomere diseases.

5. CONCLUSIONS

Short telomeres and mutations in telomerase complex genes are genetic risk factors for the development of IPF and apparently pulmonary fibrosis and emphysema syndrome. The mechanisms for disease are still. It is unclear what cells in the lungs require telomerase and long telomeres for adequate injury response and tissue regeneration, and novel cell models are necessary for the investigation of the mechanisms of the disease. Modulation of telomerase expression may be a therapeutic tool for the treatment of these disorders.

REFERENCES

1. Dokal I. Dyskeratosis congenita in all its forms. *Br J Haematol.* 2000;110:768–779.
2. Calado RT, Young NS. Telomere diseases. *New Eng J Med.* 2009;361:2353–2365.
3. Langston AA, Sanders JE, Deeg HJ, et al. Allogeneic marrow transplantation for aplastic anaemia associated with dyskeratosis congenita. *Br J Haematol.* 1996;92:758–765.
4. Yabe M, Yabe H, Hattori K, et al. Fatal interstitial pulmonary disease in a patient with dyskeratosis congenita after allogeneic bone marrow transplantation. *Bone Marrow Transplant.* 1997;19:389–392.
5. Giri N, Lee R, Faro A, et al. Lung transplantation for pulmonary fibrosis in dyskeratosis congenita: case report and systematic literature review. *BMC Blood Disord.* 2011;11:3.
6. Alder JK, Cogan JD, Brown AF, et al. Ancestral mutation in telomerase causes defects in repeat addition processivity and manifests as familial pulmonary fibrosis. *PLoS Genet.* 2011;7:e1001352.
7. Nunes H, Monnet I, Kannengiesser C, et al. Is telomeropathy the explanation for combined pulmonary fibrosis and emphysema syndrome?: report of a family with TERT mutation. *Am J Respir Crit Care Med.* 2014;189:753–754.
8. Raghu G, Collard HR, Egan JJ, et al. An official ATS/ERS/JRS/ALAT statement: idiopathic pulmonary fibrosis: evidence-based guidelines for diagnosis and management. *Am J Respir Crit Care Med.* 2011;183:788–824.
9. Kekevian A, Gershwin ME, Chang C. Diagnosis and classification of idiopathic pulmonary fibrosis. *Autoimmun Rev.* 2014;13:508–512.
10. Coultas DB, Gong Jr H, Grad R, et al. Respiratory diseases in minorities of the United States. *Am J Respir Crit Care Med.* 1994;149:S93–S131.
11. Selman M, Pardo A. Idiopathic pulmonary fibrosis: an epithelial/fibroblastic cross-talk disorder. *Respir Res.* 2002;3:3.
12. King Jr TE, Pardo A, Selman M. Idiopathic pulmonary fibrosis. *Lancet.* 2011;378:1949–1961.
13. Hodgson U, Pulkkinen V, Dixon M, et al. ELMOD2 is a candidate gene for familial idiopathic pulmonary fibrosis. *Am J Hum Genet.* 2006;79:149–154.
14. Seibold MA, Wise AL, Speer MC, et al. A common MUC5B promoter polymorphism and pulmonary fibrosis. *New Eng J Med.* 2011;364:1503–1512.
15. Thomas AQ, Lane K, Phillips 3rd J, et al. Heterozygosity for a surfactant protein C gene mutation associated with usual interstitial pneumonitis and cellular nonspecific interstitial pneumonitis in one kindred. *Am J Respir Crit Care Med.* 2002;165:1322–1328.
16. Mushiroda T, Wattanapokayakit S, Takahashi A, et al. A genome-wide association study identifies an association of a common variant in TERT with susceptibility to idiopathic pulmonary fibrosis. *J Med Genet.* 2008;45:654–656.
17. Tsakiri KD, Cronkhite JT, Kuan PJ, et al. Adult-onset pulmonary fibrosis caused by mutations in telomerase. *Proc Natl Acad Sci U S A.* 2007;104:7552–7557.
18. Armanios M, Chen JL, Chang YP, et al. Haploinsufficiency of telomerase reverse transcriptase leads to anticipation in autosomal dominant dyskeratosis congenita. *Proc Natl Acad Sci U S A.* 2005;102:15960–15964.
19. Armanios MY, Chen JJ, Cogan JD, et al. Telomerase mutations in families with idiopathic pulmonary fibrosis. *New Eng J Med.* 2007;356:1317–1326.
20. El-Chemaly S, Ziegler SG, Calado RT, et al. Natural history of pulmonary fibrosis in two subjects with the same telomerase mutation. *Chest.* 2011;139:1203–1209.
21. Armanios M. Telomerase and idiopathic pulmonary fibrosis. *Mutat Res.* 2012;730:52–58.
22. Kropski JA, Mitchell DB, Markin C, et al. A novel dyskerin (DKC1) mutation is associated with Familial Interstitial Pneumonia. *Chest.* 2014, in press. http://dx.doi.org/10.1378/chest.13-2224.
23. Alder JK, Chen JJ, Lancaster L, et al. Short telomeres are a risk factor for idiopathic pulmonary fibrosis. *Proc Natl Acad Sci U S A.* 2008;105:13051–13056.

24. Diaz de Leon A, Cronkhite JT, Katzenstein AL, et al. Telomere lengths, pulmonary fibrosis and telomerase (TERT) mutations. *PLoS One*. 2010;5:e10680.
25. Aalbers AM, Kajigaya S, van den Heuvel-Eibrink MM, van der Velden VH, Calado RT, Young NS. Human telomere disease due to disruption of the CCAAT box of the TERC promoter. *Blood*. 2012;119:3060–3063.
26. Chiang YJ, Calado RT, Hathcock KS, Lansdorp PM, Young NS, Hodes RJ. Telomere length is inherited with resetting of the telomere set-point. *Proc Natl Acad Sci U S A*. 2010;107:10148–10153.
27. Alder JK, Guo N, Kembou F, et al. Telomere length is a determinant of emphysema susceptibility. *Am J Respir Crit Care Med*. 2011;184:904–912.
28. Calado RT. Telomeres and marrow failure. *Hematology Am Soc Hematol Educ Program*. 2009;338–343.
29. Calado RT, Dumitriu B. Telomere dynamics in mice and humans. *Semin Hematol*. 2013;50:165–174.
30. Liu T, Chung MJ, Ullenbruch M, et al. Telomerase activity is required for bleomycin-induced pulmonary fibrosis in mice. *J Clin Invest*. 2007;117:3800–3809.
31. Liu T, Ullenbruch M, Young Choi Y, et al. Telomerase and telomere length in pulmonary fibrosis. *Am J Respir Cell Mol Biol*. 2013;49:260–268.
32. Bitterman PB, Rennard SI, Keogh BA, Wewers MD, Adelberg S, Crystal RG. Familial idiopathic pulmonary fibrosis. Evidence of lung inflammation in unaffected family members. *New Eng J Med*. 1986;314:1343–1347.
33. Krizhanovsky V, Yon M, Dickins RA, et al. Senescence of activated stellate cells limits liver fibrosis. *Cell*. 2008;134:657–667.
34. Rodier F, Campisi J. Four faces of cellular senescence. *J Cell Biol*. 2011;192:547–556.
35. Chilosi M, Carloni A, Rossi A, Poletti V. Premature lung aging and cellular senescence in the pathogenesis of idiopathic pulmonary fibrosis and COPD/emphysema. *Transl Res*. 2013;162:156–173.
36. Yamaguchi H, Calado RT, Ly H, et al. Mutations in TERT, the gene for telomerase reverse transcriptase, in aplastic anemia. *New Eng J Med*. 2005;352:1413–1424.
37. Fogarty PF, Yamaguchi H, Wiestner A, et al. Late presentation of dyskeratosis congenita as apparently acquired aplastic anaemia due to mutations in telomerase RNA. *Lancet*. 2003;362:1628–1630.
38. Calado RT, Brudno J, Mehta P, et al. Constitutional telomerase mutations are genetic risk factors for cirrhosis. *Hepatology*. 2011;53:1600–1607.
39. Calado RT, Regal JA, Kleiner DE, et al. A spectrum of severe familial liver disorders associate with telomerase mutations. *PLoS One*. 2009;4:e7926.
40. Agarwal S, Loh YH, McLoughlin EM, et al. Telomere elongation in induced pluripotent stem cells from dyskeratosis congenita patients. *Nature*. 2010;464:292–296.
41. Batista LF, Pech MF, Zhong FL, et al. Telomere shortening and loss of self-renewal in dyskeratosis congenita induced pluripotent stem cells. *Nature*. 2011;474:399–402.
42. Winkler T, Hong SG, Decker JE, et al. Defective telomere elongation and hematopoiesis from telomerase-mutant aplastic anemia iPSCs. *J Clin Invest*. 2013;123:1952–1963.

INDEX

Note: Page numbers followed by "*f*" indicate figures and "*t*" indicate tables.

U

Z

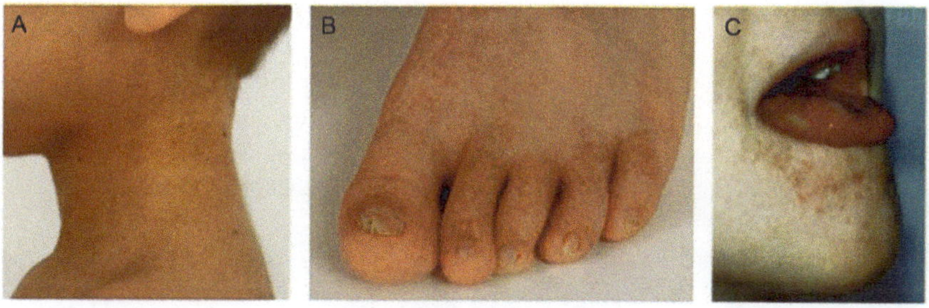

SHARON A. SAVAGE, FIGURE 2.1

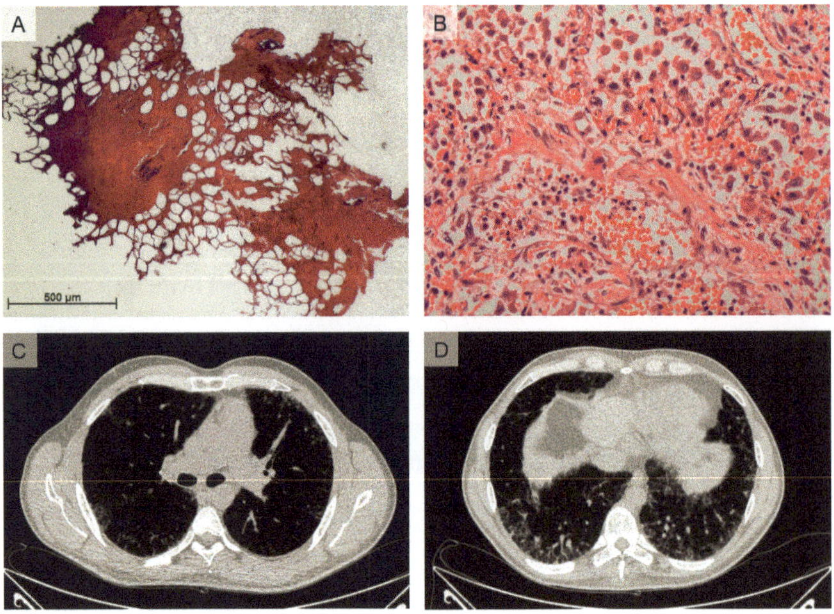

Rodrigo T. Calado, Figure 8.1

FSC
www.fsc.org
MIX
Papier | Fördert gute Waldnutzung
FSC® C083411

Kolibri 360 GmbH
Ferdinand-Jühlke-Straße 7
99095 Erfurt, Deutschland
produktsicherheit@kolibri360.de